I0475919

Origins: Before the Big Bang
An Email Conversation With 100 Scientists

Origins: Before the Big Bang
An Email Conversation With 100 Scientists

Compiled and Edited by

Ken Masters

Published by: Lulu.com
2015

Copyright © 2015 rests with the individual authors of the texts.

First Printing: 2015

ISBN (Paperback copy): 978-1-312-75326-6

Lulu.com

Dedication

To all who share their knowledge

Contents

Acknowledgements

My thanks to the scientists who responded to my queries; thanks also to those who received my original mail, but were not in a position to answer my question, and who passed my queries on to others to answer.

Preface

Why this book?

It is a bit of a cheek, I think, my producing this book. I am a layperson with very little knowledge of physics, apart from high school and a few first-year university courses that I bravely attempted with mixed results. This is not a matter of pride; it is simply a statement of fact.

Normally, my lack of qualifications in physics would be an automatic disqualification in the production of a book like this, especially given the complexity of the topic. In this case, though, it stands me in good stead. While this book is not quite in the "Explain it like I'm a 5 year old" (ELI5) category, it is aimed at the general reader, so I am a good filter for the arrangement and use of the content.

It started with a simple question that, I am sure, many people outside the fields of cosmology, physics and astronomy have asked: How did it all begin? Having had the mantra that "energy and matter cannot be created or destroyed, only transferred and transformed" drummed into me at high school, and having heard the basics of the Big Bang – that "everything" started from a massive explosion of some type – I wanted to find out exactly where that initial stuff that exploded actually originated. The obvious questions that came along with that were: why it originated in the first place, why in that particular format, and then, why did it suddenly change?

I searched several popular texts and watched documentaries, and read papers that I could understand, but the moment I began to understand something, it seemed that others had different ideas. I also heard, several times, phrases like "it is at that stage that the laws of physics break down," and this sounded to me like a "something strange happens here" box in a series of steps in a process, and is of no value at all. So, I was left back at square one.

How did this book come to be?

Anyone who knows me knows that I'm not partial to uncertainty, and so, with this uncertainty, I decided to ask the experts. And not only the famous experts, but also many people who are working around the topic, so that I could get some sort of broader picture. But who are these experts? I realised that there was no fool-proof method of determining this, so I settled on a method that, while not flawless, would probably work reasonably well.

Given that the experts were *probably* working at education and research institutions, and given that they would *probably* be working at highly ranked institutions, I started there, in March 2014.

As there are several world ranking systems of universities, I used:

- QS University Rankings
- The Times Higher Education World University Rankings
- Webometrics Ranking Web of Universities
- Academic Ranking of World Universities (ARWU)

From each of these, I gathered the top 200 ranked institutions, sorted and sifted, and then spent quite a bit of time searching through their web pages, hunting for likely representatives in the fields of astronomy, cosmology and physics.

Having obtained a list of experts, in April and May 2014, I sent out the following email:

I am writing to you as [Position] [Department] [University] to ask a simple layman's question that has been bothering me for some time: Given that energy and matter cannot be created, how was all the original matter and energy of the universe created in the first place? That is, before the Big Bang?

The reason I'm asking is that I wondered why it existed at all? And in that form? And for how long? And why did it suddenly change? (I have heard people say things like "There was nothing before the Big Bang, because time did not exist before it" or "The laws of physics were different then," or "it just is," but these sound like cop-outs to me.)

I would really appreciate your taking the time to answer my question. Should you feel, however, that this is not an area of your expertise, could you please forward this mail to a member of your staff who would be prepared to answer it.

I received more than a hundred responses. Some were short, some much longer, and several gave references to further reading. While I was working through the responses, I realised that I had in my possession something that many people would like to see. From that, came the idea to create this book.

As the emails sent to me had been written as private letters, I replied to those who had responded, informed them that I would like to create this small book of their responses, and offered them a chance to withdraw or edit their contributions, and also a chance to remove their identity and have their contributions marked as anonymous.

Based on their replies, I was able to use 100 responses. In October 2014, I sent out a final copy for approval (with this Preface and Introduction), and the result is this book.

Editing and arranging the responses

This book contains the final responses. Although I have edited the responses, as far as possible, I have attempted to retain the voice of the scientists. This was an informal conversation between experts and a lay-person, and I wished to keep it at that level. It was necessary, however, to apply small edits. Typical edits included:

- Removal of greetings.
- Small punctuation corrections or changes for ease of reading.
- Formatting (e.g. superscripting) that is not always possible in emails.
- Breaking some long paragraphs into smaller paragraphs.
- For copyright reasons, not reproducing articles sent to me, but rather supplying bibliographic details.
- Replacing links to books' (usually on vendors' pages) bibliographic information as a reference. (Links, where available, are supplied in the bibliographic information.)

- Removing some extraneous information about others in their department or institution, unless the conversation of one linked directly to the other.

I had considered arranging and grouping the responses according to some criteria, either by country, or by name of scientist, or by opinion. But any grouping arrangement I chose would run the risk of shaping the reader's experience, and I felt had no right to do that. Therefore, I have left the responses in the order in which the scientists initially responded to my queries.

Some comments

Before ending this preface, I shall take some liberty in giving my comments on the responses.

Firstly, a disappointment. A worrying issue that raised itself several times in the initial responses to my question was my religion: several scientists wanted to know more about me and my religious views before answering my question. Although they asked politely, and I responded likewise, I did find it disconcerting. Yes, I am aware of religion as part of the debate, but that's not an explanation. I have been a student and teacher for more than 30 years, and I cannot remember ever asking or being asked that question, and having the gate to knowledge closed to me, and opened only on condition that my religious affiliation and opinions were first declared. As most of these scientists are teachers, I did wonder if their students are also required to declare their religion before taking their classes, or accessing the libraries at these institutions.

I have heard of these attitudes in "closed" societies, and amongst people who are commonly referred to as backward, or prejudiced; I had not expected to find it in 21st century scientific circles. I wondered what would have happened if people like Newton or Lemaître had faced this early in their careers.

Correspondingly, however, it turns out that several scientists have been asked to engage in innocent conversations, only to find their material misused by those with a secret agenda. So, some of the responsibility for this reluctance to engage lies with those who would abuse the good intentions of those scientists.

Secondly: a general attitude of sharing. Notwithstanding my initial comment, the responses were welcomingly open. As expected, people were busy, and did not always have time to respond in detail. But many took the time to explain a great deal, and many were able to supply me with useful references; in some cases, even with copies of articles that were not easily available.

Thirdly: the timing. My mails and most of the responses were sent shortly after the BICEP2 publication, and before the results were questioned, so several of the responses refer directly to those results, or to the optimism of those results. They need to be viewed in that light.

Fourthly: the wide range of opinions. The scientists in these conversations draw on a wide range of theories, such as Quantum Theory, General and Special Relativity, Multiverses and other concepts. All have different levels of certainty, limitations and of applicability. As a result, the scientists themselves have different levels of certainty in their answers, and they range the full spectrum from absolute certainty to hypotheses to speculation.

While the reader might become frustrated at this range, I believe that it is precisely this range of opinions and the strength of conviction that gives this book a distinctive and valuable character. While many popular books on the topic have been written, they tend to give the authors' opinion, with a passing reference to other ideas. This book gives no special place to any opinion or argument. My own beliefs have not influenced the collection in any way.

So, after all that, I'm probably not much closer to my answer. But, at least I know I'm not alone.

Introduction

As I am not an expert in this field, I shall keep the Introduction to this book short.

From my readings of the letters and the references contained in this text, I can see that it is obvious that the origin of the universe is something that has been debated for almost as long as humans have existed. Various philosophical and scientific texts have explored the questions posed, and a wide range of opinions have been given.

The text highlights many of the tensions between theoretical and experimental scientists. In addition, we see that, for some scientists, the answers are clear, with, perhaps, some refinements to details required; others are less certain, and some suggest that all hypotheses on the topic are speculation.

In one way, nothing has changed in the past several thousands of years. Greek philosophers struggled with uncertainties, wondering if the limits lay within themselves, or the tools at their disposal, or the nature of the universe itself. Similarly, there were those who believed in the certainty of their craft, and that the few details that had to be resolved were trifles.

There is an ironic lesson to be learnt here. Certainly, in the past few thousand years, our knowledge of the universe has increased dramatically. Academic fields have changed, expanded, and new fields developed. And yet, we still have the full continuum of thinkers, from those who question what we know, and wonder if we will ever be able to know the real truth, to those who are certain that we have the answers, and are merely on the brink of ironing out the minor details. In my mind, they echo somewhat Phillip von Jolly's pronouncement at the end of the 19th century that everything in physics had already been discovered, and "all that remains is a few holes," which, in turn, echoes those writings of Ptolemy, sketching out the complete universe, with just a few minor observations that don't quite fit the picture.

Is the first group realistically circumspect, or merely too timid to trust their science? Alternately, is the second group standing on solid science, or a little too sure of themselves and a little oblivious to the number of times humans have said the same thing in the past, ignoring the possibility that tomorrow may prove our current thought incorrect?

Either way, the journey through these pages promises to be interesting, and I hope you enjoy the ride.

100 responses to my mail

We can only see a short distance ahead, but we can see plenty there that needs to be done. (Alan Turing, 1959)

Professor Leif Lönnblad
Head: Department of Astronomy and Theoretical Physics
Lund University
Sweden

They are indeed cop-outs. In physics we deal with mathematical explanations and descriptions of observed phenomena. We have wonderful, extremely well tested, theories explaining almost everything we see, but the problem is that they are not internally consistent when they are extrapolated to the moment of the Big Bang.

There are several suggestions for new theories that may be consistent (such as Super String Theory), and could say something about the Universe even before the Big Bang, but these have not been confirmed by any observations. Therefore, to my mind, the only reasonable answer to your question is "we simply don't know."

Regards,

---ooOoo---

Professor Csaba Balazs
Director: Monash University Node
Centre of Excellence for Particle Physics
Australia

You are asking one of the big questions, and the short answer is: we don't know. Nobody can "correctly" answer your main question, because, today, physics is exploring processes at energies of 1 Tera electron Volts (TeV), while the Big Bang probably happened around 1,000,000,000,000,000 TeV. Simply put, we cannot see as far ("in energy") as the Big Bang, based on our present insight.

You can imagine the answer as a big puzzle eventually coming together. As always in physics, we already have a few pieces of this puzzle. But anyone who claims to know what happened around the Big Bang is trying to mentally complete a puzzle with thousands of pieces missing. I don't attempt to do that. I'm working on puzzles that have a chance to be completed in my life-time.

PS. You can (if you haven't yet) start reading http://en.wikipedia.org/wiki/Big_Bang from which you can follow links to millions of other pages (and then you begin to appreciate the number of pieces of the puzzle).

Cheers,

---ooOoo---

Professor Tim Greenshaw
Head: Department of Physics
University of Liverpool
UK

We don't really know the answer to this question, but here are some brief comments!

There isn't a "before" the Big Bang. Time and space were created at the Big Bang, not just matter/energy. The Big Bang doesn't involve something expanding into a pre-existing space and time; space and time are created in the Big Bang as well as matter/energy.

The most common picture of how the Universe began is that time, space and matter/energy were all brought into existence in a quantum fluctuation. (Heisenberg's Uncertainty Principle says you can create energy - violating energy conservation - as long as it only exists for an extremely short time.) Because of the nature of the matter/energy created, this fluctuation was blown up from being absolutely miniscule to something like a football sized "Universe" in a tiny fraction of a second, and this prevented the quantum fluctuation from just vanishing again, as would normally be expected.

From this point, the particles and forces we now see started to develop (quarks, electrons etc.)

I can't really say more in a short email...

All the best,

---ooOoo---

Professor Thomas Killian
Chair: Department of Physics and Astronomy
Rice University
USA

There are many books written for a popular audience that might give you an introduction, such as The Origin of the Universe (Barrow 1997). This is an open area of research and many questions remain unanswered.

Regards,

---ooOoo---

Professor Ewan Stewart
Theoretical High Energy Physics Group
Department of Physics
Korea Advanced Institute of Science and Technology (KAIST)
South Korea

It is an important question that motivates Inflation. You should be able to find lots of information about Inflation on the web, including my homepage (http://profstewart.org/).

---ooOoo---

Professor Paul Davies
Director: Beyond Center for Fundamental Concepts in Science
Arizona State University
USA

All of these commonly asked questions have straightforward answers, and I have answered them all, in detail, in easy language, in my book *The Goldilocks Enigma* (Davies 2008). Please excuse me if I refer you to the book rather than attempt long summaries here.

Happy reading!

---ooOoo---

Anonymous
Switzerland

In fact, there is no good physics answer to this question. When we extrapolate the evolution of the Universe back in time, we arrive at the Big Bang where our equations simply break down. This means that, with our current theory, we cannot make any statements about the Universe when we approach the Big Bang, and, of course, also not answer the question about what happened before it.

Sorry if this answer is somewhat unsatisfactory.

Regards,

---ooOoo---

Professor George Smoot
Teacher-researcher
U.F.R. Physique
Université Paris Diderot - Paris 7
France

It is thought that the total energy inside our observable portion of the Universe is zero (0) or slightly negative. The total content minus the gravitational binding energy is about zero to high order. So, there was not necessarily any energy created or destroyed in the creation of the Universe.

---ooOoo---

Origins: Before the Big Bang

Professor Alexis Brandeker
Stockholm Observatory
AlbaNova University Center
Sweden

No-one knows for sure, but there are two possibilities:

1. Taking potential energy into account (which can be negative), the total amount of energy in the Universe could actually be exactly zero.

   See:
   http://machineslikeus.com/news/big-bang-beginners-13-does-big-bang-theory-violate-law-conservation-energy
   http://www.newton.dep.anl.gov/askasci/phy05/phy05329.htm

2. Conservation of energy was violated.

These are not easy questions, since, at very early times, the Universe was small enough that both gravity and quantum effects were both important, and, unfortunately, there is no complete theory for this yet. Moreover, it is not known how far the laws of physics can be extrapolated to earlier times. There has recently been a tentative break-through in this respect, the BICEP2 discovery of ripples in the cosmic microwave background that confirm predictions about the Universe at the age of 10^{-36} seconds.

You can read more about the Big Bang and common questions related to it at

http://www.cfa.harvard.edu/seuforum/faq.htm
http://science.howstuffworks.com/dictionary/astronomy-terms/big-bang-theory.htm

Yours sincerely,

---ooOoo---

Origins: Before the Big Bang

Professor Peter Quinn
ICRAR Director
Astronomy and Astrophysics
University of Western Australia
Australia

The questions you are asking are not easy to answer, because we really do not know the answers, and we actually do not have good tools or ideas to look for the answers. This makes them great questions. They are the kind of questions that drives knowledge and understanding forward. Let me try to add a few ideas that may help.

The work of Einstein in the early part of the 20th century showed us that matter and energy are really two sides of the same thing. I can turn matter into energy, and I can make matter from energy. This is the famous $E=mc^2$ equation, where E is energy, m is mass and c is the speed of light. So, if you have enough energy in one place, like the Big Bang, then you can create a lot of mass.

So, the question is, then, where did all the energy come from? Did it exist before the Big Bang? Did something else create the energy and then the Big Bang happened?

The problem we face is the one that you pointed to. We do not have any physical laws that we can use to describe the Big Bang itself. Our two best theories, the theory of gravity and the quantum theory of matter, both fail when you try to use them to describe the Big Bang. They work fantastically well when you apply them at any other time in the history of the Universe. This lack of a good theory to talk about the Big Bang is probably the biggest problem in all of physics right now. I think we are waiting for the next Einstein to come along with the right idea to crack this.

What we do know is that time and space are connected (again courtesy of Einstein's work on gravity). So, if there is no space (before the Big Bang), then there is no time. That means it's very hard to think about the pre- Big Bang Universe, because we have no notion of how things change, i.e. there is no time, as such.

Some people have proposed that the Universe actually does not have a Big Bang which is a singularity (an infinitesimal point), but it actually stops and "bounces" before it gets to zero size. Maybe, then, the Universe is bouncing between Big Bangs and Big Crunches. In that case, the energy and mass would be constant. It still doesn't account for the origin of the Universe in the first place, but is consistent with a Universe that has always been there.

Ken, these are good and interesting questions. Ones I (and maybe all physicists) do not have all the answers for. That's what makes them fun to think about!

Regards

---ooOoo---

Professor Nicolás Cardiel
Departamento de Astrofísica
Facultad de Ciencias Físicas
Universidad Complutense de Madrid
Spain

The answer is simply that matter and energy can be created from a vacuum, given the right conditions. I refer you to the article "The Universe: the ultimate free lunch" (Stenger 1990) that explains this idea.

I hope this can help you understand how Physics is trying to understand how the Universe was created.

Best regards.

---ooOoo---

Professor Cecilia Woo
Department of Applied Physics
Hong Kong Polytechnic University
China

Thanks a lot for the interesting question. I will try to answer it even though I'm not a true expert on cosmology.

You are right that energy-matter cannot be created. Therefore, in the theory of modern cosmology, all the matter/energy is assumed to have already existed at the beginning. Our current knowledge of physics is, unfortunately, not applicable to the singularity at the Big Bang, so, strictly speaking, we do not know anything before that, nor we have definite answers to important questions like the origin of matter and what triggered the Big Bang. That is why some people consider these as philosophical or metaphysics questions.

Of course, one may apply the anthropic principle to claim that, should there be no Big Bang, humans would not have existed to ask such questions. However, that still does not answer your question.

One model that was somewhat popular before was the cyclic Universe, which suggests that the expansion of the Universe may eventually slow down due to gravity. The Universe will then collapse and a Big Bang (preceded by a Big Crunch) would occur again. This cycle repeats, and the Universe oscillates forever. But, after the discovery of the accelerating Universe (Nobel Prize in physics 2011), the oscillating Universe model becomes less favourable, as our Universe now seems to expand faster and faster, and gravity is not enough to pull it back.

Finally, there are some alternative theories to Big Bang, such as loop quantum gravity, but I am not too familiar with them.

I hope you will find this information useful.

Best,

---ooOoo---

Professor Nial Tanvir
Department of Physics and Astronomy
University of Leicester
UK

You have obviously already done some reading on this subject, I'll see if I can clarify a few points, but bear in mind that these are questions that we are far from having full answers to! So:

1. The "conventional" theoretical description of the Big Bang is based on Einstein's General Theory of Relativity (GR), and, in this description, the mathematics indeed tells us that time (and space) started then: the equations to not extend back to any prior time. The reason is that, in this picture, the Universe starts with a "singularity," which is a point of infinite density and temperature.

2. Now, even in this picture, it is not obvious that creation of matter and energy is forbidden - in part, because the singularity itself is not described by the theory, so it may be "allowed" to create matter and energy. Also, it is possible that one can side-step the question if the total energy is zero: since potential energy is negative, it might cancel out all of the positive forms of energy.

3. However, for various reasons, particularly associated with quantum mechanics, it is widely believed that the GR description can only be an approximation. For decades, people have been working on trying to develop an improved mathematical theory, and some progress has been made, but we still don't have one (it would generically be called quantum gravity, and the most promising scheme is usually said to be String Theory, but as I say, we still don't have any fully worked out replacement for GR).

   Anyway, the hope/expectation is that, if we can develop a successor to GR, it will do away with the singularity and would provide a complete description of the Big Bang, possibly answering the question of where it all came from.

   However, it is worth bearing in mind a few final points:

4. The mathematical description of the Big Bang that we do have does seem to be very good. It allows us to say, with some confidence, what was happening in the Universe less than a trillionth of a second after the start - at this point, the particle energies were comparable to those achieved at CERN, so the physics is fairly well understood. Most cosmologists believe that the theory probably works well at much shorter times still, possibly as early as a trillionth of a trillionth of a trillionth of a second (but not much before that).

5. If, and when, a quantum theory of gravity is developed, it is (unfortunately) unlikely to give a picture of the Big Bang that is easy to visualise – i.e. as you say, we could imagine the Universe sitting around in some state, and then suddenly ex-

ploding, but the progress we have already made in developing quantum gravity suggests a much less intuitive picture, for example involving many extra dimensions beyond the three we perceive.

Well, I hope that has given you some further food for thought.
Regards,

---ooOoo---

Professor Dr. Viatcheslav Mukhanov
Theoretical Astroparticle Physics and Cosmology
Ludwig-Maximilians-Universität München
Germany

A gravitational field has negative energy. Therefore, you can start with zero energy and produce a lot of matter with large positive energy which will be compensated by negative energy of gravitational self-interaction. That way, you will get a lot of matter, the total energy of which will still remain zero. Gravity is an infinite reservoir from which you can borrow energy for producing matter.
Best regards

---ooOoo---

Dr. Gary Mathlin
Department of Physics
University of Bath
UK

The question you ask is basically still open and the subject of much ongoing theoretical work, but I'll outline one way of thinking about it - but not the only way - below.

The question of what happened before the Big Bang may actually be poorly founded. Being creatures of metre scale, able to propel ourselves to velocities of the order of metres per second, we perceive space, the distance between things, and time, the sequencing of events, as separate entities. However this is just an illusion. Space and time do not exist separately; all that really exists is space-time. Furthermore, in our naive view, space and time appear to be immutable, forming a backdrop for events in the Universe to play out against. In reality, space-time is malleable, and plays an active role in the Universe. This all sounds highly theoretical, but the everyday technology of sat-nav only works because we know how to account for the warp in space-time caused by the mass of the Earth.

If you imagine playing the "video" of the expanding Universe backwards, so everything gets closer together, we see the density of the Universe increase, and, thus, space-time becoming more and more curved. As the density approaches infinity, the temporal part of space-time becomes curved in on itself. As we approach the Big Bang itself, the temporal component can only point in one direction, to the future. If you travel to the North Pole on Earth, direction becomes meaningless: the only direction you can face at the North Pole is south - there is no east, west or north. Asking what came before the Big Bang is like asking what's north of the North Pole.

Of course, this doesn't address the issue of where all the matter and energy came from. One way to address this is to appeal to quantum physics, and the "Heisenberg Uncertainty Principle," in particular. Although this doesn't violate the principle of the conservation of matter, it does allow for energy (or matter which is really the same thing) to pop into existence out of nothing. The amount of energy that "appears" is inversely proportional to the amount of time that it can exist for. This phenomenon, although weird sounding, is observed in the laboratory routinely. It is possible that the Universe came into existence as a result of a random quantum fluctuation.

The above is a very brief, off the cuff, response to your questions and is certainly not the definitive answer. If you wish to explore this further, I recommend having a look at an interesting idea by Roger Penrose called "conformal cyclic cosmology" which he wrote up for the general reader as Cycles of Time (Penrose 2012).

I hope my reply is of some use to you.

Best wishes

---ooOoo---

Professor Steve Lloyd
Head: School of Physics and Astronomy
Queen Mary, University of London
UK

An interesting question, of course, to which no-one knows the answer.

Some people believe there may have been previous Universes, but that only postpones the question. One theory, which I particularly like, but have no idea how plausible it really is, is that the total energy of the Universe is zero. Obviously the matter (stars, us etc.) has positive energy, but a gravitational field has negative energy, as you have to put energy in to separate two objects. Hence, it could be possible that the positive energy of the matter is balanced by the negative potential energy of gravity, and the total is zero.

Where did the Universe come from? – well, presumably a quantum fluctuation of the (zero energy) vacuum or whatever - just like quantum fluctuations can produce particle anti-particle pairs from the vacuum. Not sure if this helps - I'm a Particle Physicist not a Cosmologist!

Best regards

---ooOoo---

Dr. John Gribbin
Department of Physics and Astronomy
University of Sussex
UK

This may help: "Everything from nothing," http://johngribbinscience.wordpress.com/2013/02/17/everything-from-nothing/

---ooOoo---

Professor Andreas Reisenegger
Director: Astronomy & Astrophysics
Pontificia Universidad Católica de Chile
Chile

The questions you ask are at, or even beyond, the current frontiers of science.

We think we understand what happened only a small fraction of a second after the Big Bang, but not at or even less before that, if there was a "before" at all. We'll keep looking! :-)

Best regards,

---ooOoo---

Professor James Beatty
Chair: Department of Physics
Ohio State University
USA

The Big Bang occurs at a singularity in space-time. A simple analogy with some of the features of this is the way coordinates on the surface of the earth behave at the Poles. If you start at the North Pole, every direction is south, and these directions are east and west of each other as soon as you get a tiny distance from the North Pole. At the Pole there is no north, east, or west, just south.

At the Big Bang singularity, space itself shrinks to a point so there is no x, y, or z coordinate. Time only exists forward, so the word "before" has no meaning here.

There are some models for why this might be the case, but I'd class these as pure speculation with no observable es. Most of these involve our Universe being something analogous to a bubble in a higher dimensional space-time.

---ooOoo---

Professor Rob de Ruyter
Chair: Department of Physics
Indiana University Bloomington
USA

Thank you for your questions about cosmology and the Big Bang.

The questions you are asking are profound, and at the edge of current research. There are ideas on all of them, but I do not think that there is clear consensus on all. I wish I was able to give you a short answer in an email, but I hope you appreciate that that is not really possible: I, myself, am not in this field, but more importantly, to get an appreciation for the current thinking on these topics requires a long period of training in mathematics and physics.

An email would certainly not do this justice. I think the best I can do is to point you to a few recent books that deal with various aspects of these questions (see below). Some of the ideas treated there are quite abstract, and not easy to understand at first. They require work and thought to sink in. So I would recommend that you take the time to read and think about these issues. I hope this helps.

Lawrence Krauss: *A Universe from Nothing* (Krauss 2013). This book comes closest to answering your specific question about where the Universe comes from, and also answers, to some extent, your feeling that some answers are "cop-outs."

L. Lederman & C. Hill: *Symmetry and the Beautiful Universe* (Lederman & Hill 2007). This is a text that treats the relations between mathematical symmetry and the structure of the world around us.

T. Duncan and C Tyler: *Your Cosmic Context* (Duncan 2008). This textbook presents a good introduction, without a great deal of mathematics.

Sincerely,

---ooOoo---

Origins: Before the Big Bang

Professor Samuel Matteson
Department of Physics
University of North Texas
USA

Your question is a profound one. Many things about the origins of the physical world are counter-intuitive precisely because our instinctive ideas are formed from our day-to-day experiences. But it appears from the evidence that the conditions were far, very far from ordinary at the origin of the Universe.

The data show that almost all of the Universe is rushing away from all the rest (except for a few objects "nearby" us). Looking back in time as we look with telescopes, we see the objects (galaxies) moving away even faster than they are now. If we "play the motion picture" backward (speaking metaphorically), we conclude that all of matter and energy and space (and since Einstein showed that space and time are conjoined) time, too, were bound up in a small package of incredibly tiny dimensions approximately 13.8 billion years ago. The density of energy and matter many, many times exceeded that we have so far produced in the most powerful accelerators.

The challenge to our imaginations is to visualize what is "outside" of space and time. This is an impossible question, or even a "non-question," like "What is the sound of purple." While it may seem a "cop-out" to say this, it is actually the case. We can better understand the situation by imagining two dimensional creatures plus time (as was done in the book *Flatland*) living on the surface of a ball-like sphere. All that they can experience is on the surface. As the sphere expands, they see everything moving away from every other thing, but it is because their space is expanding. We, being of four dimensions (3 space plus time), see things a little differently than they. Of course, we share a sense of time as do they. What would it be like to be both in time and out of it simultaneously? I am not being ironic on purpose, but I am just illustrating how trapped in our temporal perspective we are.

But how can we know what is happening or what did happen? Only by looking at what is here now and inferring what occurred in the past. It appears that, at the Big Bang (a name given to the event to disparage it by its original detractors as too incredible) all matter, all energy, space AND time itself (notice I use the singular, for space and time are one thing) began, according to most analyses. This contention so unnerves some physicists that they are hard at work trying to conjecture alternatives, but, so far, no viable theory has come forth. The problem is that, without time, there is no "before," and, without time, process ceases to be a viable concept, and we move out of the realm of physics into . . . what ? Philosophy.

The challenge for cosmologists is to attempt to explain how what we find here and now came to be by processes that we can test, for example, by predicting the distribution of the Cosmic Background Radiation and its polarization, which seems to be accurately pictured by the Big Bang theory. There are many incredible and awe-inspiring wonders in this world that continue to amaze and delight me. But the evidence is compelling:

1. This Universe, while big and old, had a beginning, i.e. there was a point of origin with no "before."
2. This is a one-off, and it will not repeat, i.e. it looks like the cosmos will keep on expanding forever.
3. There is at least one race of intelligent life in the Universe, as unlikely as it seems to find even one.

These empirical findings, while not the origin of my theism, give me confidence that my belief in God is a warranted persuasion.

Thanks again for your query. Sorry that I do not have a better explanation, but that is the state of things as we find them.

All the best,

---ooOoo---

Professor Andrew Hynes
Chair: Earth and Planetary Sciences
McGill University
Canada

It does sound like a cop-out, I agree. Unfortunately, however, that is essentially the truth. That is, when one goes back into the condensed phase as the Big Bang is approached, the laws of physics as we know it simply do not apply, so anything that one might try to say about what was going on amounts to nothing more than guesswork.

An analogy I could draw to illustrate the problem in part, is if we were to try to make predictions about what would happen in a system in which bodies were travelling at near the speed of light if we had only Newton's laws of physics (so-called 'classical physics'). We would get everything completely wrong, since we would not have incorporated relativistic effects (i.e., Einstein's insights). It is, however, in some senses, worse than that, and in some senses better, because, in the analogy I have used, we would have thought we knew what would happen (and we would have been wrong), whereas as we approach the Big Bang, many of the parameters we are dealing with tend towards infinity, and we know that the physics we have does not work under those conditions (so at least we know the physics we have would not apply).

Perhaps this clarifies things a little - I hope so...

Best wishes,

---ooOoo---

Anonymous
USA

Thank you for your interest in physics!

I am not an expert on the questions you are asking, and I am not even sure experimental physics can answer the questions you are asking, since we can only observe the Universe that we have. There are recent beautiful observations on the cosmic microwave background (CMB) that provide information on the very early Universe including the period of fast expansion known as inflation. Have a look at recent press on the CMB, and that will give you an idea of the state of study of the beginning of the Universe...but not really any information on what came before.

Sincerely,

---ooOoo---

Dr. Tanja Bode
Division of Theoretical Astrophysics
Eberhard Karls Universität Tübingen
Germany

The Big Bang, like the centres of black holes, are singularities - infinities popping up in relations - in a theory of physics.

From experience in other theories of physics, singularities are mathematical warning signs that that particular way of describing the Universe breaks down in a particular point in space or time. While observations show black holes exist, and Big Bang cosmology theories are wonderfully successful in explaining the early Universe, the precise point of the Big Bang, like the central "tear" in space-time of a black hole, are both places where something fundamentally doesn't fit together.

Exactly what needs to be changed in our approach to handle these singularities is an active area of research, with new theories being built from the ground up to handle describing such incredible energy densities as found leading up to these singularities. We don't know the solution yet, but it's likely to be some truly fundamental change which leads to people giving these "cop-out" responses which you list.

One such possible theory, loop quantum cosmology, can be evolved backwards to show the Universe bouncing back before the Universe becomes too small. The theory, unfortunately, still has problems. It's hardly in a position to claim itself an answer yet, but people are working on these and other possibilities.

Perhaps reading Stephen Hawking's *A Briefer History of Time* (Hawking & Mlodinow 2008) would be of some help. I also recommend his lecture, *The Beginning of Time*, on his website: http://www.hawking.org.uk/the-beginning-of-time.html

Best regards,

---ooOoo---

Professor Ken Lang
Department of Physics and Astronomy
Tufts University
USA

Those who speculate about what was before the Big Bang do so in mathematics that almost no one, including me, can decipher.

Moreover, they predict multiverses, so that whatever was before keeps on creating bubbles and bubbles, from a situation in which there was no matter, no energy, just nothing.

The difficulty with this scenario is that it makes no predictions; the hypothetical multiverses cannot be observed or communicated with, so you have to take their existence on faith and the idea is decidedly non-scientific. But there is nothing wrong with faith, let's call it God and let it go at that.

Best,

---ooOoo---

Professor Anthony Aguirre
Physics Department
University of California Santa Cruz
USA

I attach an article (Aguirre 2013) I wrote a while back that I think explicitly addresses a number of your questions. See if it helps.

Cheers,

---ooOoo---

Professor Charles Alcock
Director: Harvard-Smithsonian Center for Astrophysics
Harvard University
USA

You ask profound questions, to which the answer is "we don't know." We may never know.

The people who think the most deeply about this matter are the originators of the inflation theory, most notable Alan Guth at MIT and Andre Linde at Stanford. You might ask one of them, but be aware that they are exceptionally busy in the wake of the reports from the BICEP2 experiment.

Regards,

---ooOoo---

Professor Hanno H. Weitering
Professor and Head: Department of Physics and Astronomy
University of Tennessee – Knoxville
USA

Physicists attempt to explain the workings of the Universe since its early creation almost 14 billion years ago, using the laws of physics. The laws of physics are derived from experimental observations made within our Universe.

While physics has been very successful in explaining the physical world around us, we do not know why the laws of physics are as they are. We only know their consequences and how to use them. If there were an alternate Universe, the laws of physics might be different there. We simply do not know because we cannot go there and measure. Neither can we say anything about the times before the Big Bang, since our Universe did not exist at that time. In fact, because our Universe did not exist, one could argue time did not exist since space and time are interwoven according to Einstein's theory of General Relativity (which has so far survived all experimental scrutiny!!).

In light of all this, it should come as no surprise that we also cannot answer the question why there was a Big Bang.

In short, physics can only address questions that relate to our own Universe since its early creation (i.e., Big Bang).

Hope this helps.

---ooOoo---

Anonymous
USA

In my limited knowledge of this, there is no information available from before the Big Bang. The picture that I have is that if one melted down a Chevy and a Mercedes, and then looked at the atoms afterward, one could not tell if a given atom was from one car or another. Analogous to this, the information from before the Big Bang has been lost, so we can speculate, but not test ideas.

This may not be helpful, but it is what I know.

Regards,

---ooOoo---

Dr. Professor Michael Meyer
Head: Institute of Astronomy
Swiss Federal Institute of Technology Zurich
Switzerland

I am not an expert on the Early Universe, but I can give my own understanding to answer your question. Any further discussion is not fruitful on my part, so I would refer to you others to continue.

In some sense, your questions transcend science into the realm of philosophy. Our understanding of physics breaks-down at times before the Planck Time ($< 10^{-44}$ seconds after the Big Bang, where the Big Bang is an event in space-time as a singularity; in point of fact, this term is a misnomer in that we can only speculate on what happened before inflation and certainly not before the Planck Time relative to a very hot very dense state we call the Big Bang).

See George Musser's *Scientific American* blog for more (Musser 2014).

So, I cannot comment on what came before this time as a scientist, but I can speculate along with anyone as a philosophy. These are not cop-outs, but the only consistent positions that scientists can take within the scientific method as allowed by our current understanding and working framework (paradigm). Even asking the question may not be scientific in the sense that no answer can be ruled out (that makes it not an hypothesis).

That said, some scientists work on theories that have implications on what came before this 10^{-44} seconds, and those theories may have testable predications in other realms. If empirical data are such that these theories cannot be ruled out, the theories are viable. However, that does not "prove" the implications about the earliest moments of the Universe are correct, or that predictions concerning conditions and events that may be considered to "predate" the Big Bang are accurate. However, some scientists promote the idea of an oscillating Universe that is cyclic (see (Veneziano 2004)).

Other ideas are out there, but I am not an expert.

See an article by Paul Steinhardt, from Princeton (Steinhardt 2011) on Inflation, which may be useful.

I note that recent observations suggest the presence of the signature of gravitational waves in the cosmic microwave background, supporting our ideas about inflation. This discovery is referred to in the blog link I mention above (Musser 2014). The result is part of the discussion in the other two articles I reference above as well.

Best wishes.

---ooOoo---

Professor Simon Driver
School of Physics & Astronomy
University of St Andrews
UK

No answer to this one I'm afraid. It's part of that first zillionth of a second where the abundant energy cannot really be explained. Different people will give different suggestions, but there is no hard evidence or single convincing theory.

Some like the Oscillating Universe, which was popular in the 1980s, and revised recently by Penrose (i.e., the energy of the Universe came from the Big Crunch of a previous Universe.)

Penrose is, of course, the legendary Prof Roger Penrose and his book *Cycles of Time* (Penrose 2012; Wikipedia 2014d) expounding Conformal cyclic cosmology (Wikipedia 2014c) which I associate (liberally) to the earlier Oscillating Universe (Wikipedia 2014e) championed by Albert Einstein and later Richard Tolman.

The Multiverse, of which I first heard via Sir Martin Rees, is also popular. As is the idea that our entire Universe is actually embedded in a bigger Universe, and is just an excess of entropy in that larger Universe. This was proposed by Nobel Laureate Prof Tony Leggett, who gave the Nobel Lecture at St Andrews University on "Why can't time run backwards" (University of St Andrews 2009). His talk included comments on whether the entire Universe is an entropy perturbation embedded in something larger.

Sorry to not be more help, though is usually the first question we get asked for which there is no solid answer.

Regards,

---ooOoo---

Professor Mark Alford
Chair: Department of Physics
Washington University in St Louis
USA

Unfortunately, I cannot get in to a deep discussion of this topic with you. There are many places online such as http://www.physicsforums.com/ or http://www.thenakedscientists.com where you can find people who are able to engage you at greater length.

However, I will tell you my view of this.

As you go back further in time, our knowledge becomes less secure. We can extrapolate back using the laws of physics that are currently known to us, and we find a singularity where the temperature and density of the Universe goes to infinity. But that may be an artifact of our extrapolation: maybe we don't have exactly the right laws of physics. We know that the early Universe was hot and dense, but whether it "appeared from nothing" or developed from something else, we don't know.

For further information, please try some of the online resources I mentioned above.

---ooOoo---

Professor Thomas Troland
Department of Physics and Astronomy
University of Kentucky
USA

No one really knows the answers to your questions. So, there are various ideas that are very speculative at this point. Science, after all, is always incomplete in its understanding of the Universe. I suggest you take a look at the Wikipedia entry for the Big Bang (Wikipedia 2014b). Also, take a look at this article in *Discover Magazine* (Vilenkin. Alexander 2013).

Fortunately for science, there are always unanswered questions out there to work on. Your questions are among them. Who knows when (or if) we will ever know the answers?

Sincerely,

---ooOoo---

Prof. Dr. Heinz Clement
Institute of Physics
Eberhard Karls Universität Tübingen
Germany

It's a good question, but I don't have a really satisfying answer.

One of the thoughts, one might have about the Big Bang and, possibly, the time before, has to do with quantum physics. According to the latter, there are always fluctuations in the various variables (Heisenberg's Uncertainty). So, there might have been - in principle - spatially distinct quantum bubbles with positive and negative energy content at a certain instance. From these bubbles, then, parallel Universes developed, one of which happens to be our Universe.

Is this understandable?

Best regards

---ooOoo---

Anonymous
USA

The question does not take a simple answer, but, in a nutshell, you are converting gravitational energy into massive particles and kinetic energy of motion, so there is conservation of total mass-energy. For a simple description, see Alan Guth, *The Inflationary Universe*. (Guth 1998)

For possible interpretations of mass-energy conservation if one considers also times before a Big Bang, see Martin Bojowald, *Once Before Time: A Whole Story of the Universe* (Bojowald 2010).

Yours

---ooOoo---

Professor Michael Barlow
Head: UCL Astrophysics Group
UCL
USA

Thanks for your enquiry. I'm not a cosmologist myself but an astrophysicist, i.e. I deal with the baryonic component of the Universe. However, I think it's fair to respond to your enquiry by saying that current physics is unable to say what might have existed before the Big Bang took place. There are, of course theories, such as multiverse models or String Theories, that have attempted to hypothesise earlier conditions. However, until any of these models make unique predictions that are susceptible to experimental verification, hypotheses are all that they will continue to be.

On the question of "what came before the Big Bang?", my own rationalisation is that just as one cannot ever reach a temperature corresponding to absolute zero (degrees Kelvin), but instead one can only approach it asymptotically, i.e. getting to 0.1 K, then 0.01 K, then 0.001 K, and so on (see the page "World Record in Low Temperatures" at:
http://ltl.tkk.fi/wiki/LTL/World_record_in_low_temperatures and the Wikipedia page on "Absolute Zero" (Wikipedia 2014a), then, similarly, one can never reach time T=0, but only approach it asymptotically,

in powers of ten. So, if today is T=1.3 x10^{10} yrs, then one can go back in successive powers of ten, reaching T=1 yr, ... T=0.1 sec, 0.01 sec, 0.001 sec, ... 1 x 10^{-30} sec, 1 x 10^{-31} sec, and so on, with each successive power of ten corresponding to its own era in the history of the Universe.

Best wishes,

---ooOoo---

Professor Graham Kribs
Department of Physics
University of Oregon
USA

I can interpret your question/comment "energy and matter cannot be created" in two ways, so let me answer both ways:

Matter and energy can be created, in the sense that energy can be converted to matter, and vice versa. Energy-momentum conservation means there is no free lunch, so there has to be some energy or matter to be able to go back and forth.

You may be asking a different question, which is, "What in the early Universe provided the 'initial' burst of energy that led to all of the matter and energy we observe today?" This we don't know. The current thinking, consistent with observations in the CMB, is that, at the end of inflation, the particle responsible for inflation decayed and that caused the Universe to heat up to a very high temperature (with all types of matter and energy created). As the Universe cooled, heavy particles "froze out" in the Universe, and if they were unstable, they decayed into lighter species.

I hope this helps,

---ooOoo---

Origins: Before the Big Bang

Professor Terry Herter
Chairman: Department of Astronomy
Cornell University
USA

You will find some interesting discussions regarding the Big Bang at the "Ask an Astronomer" web site at Cornell. See: http://curious.astro.cornell.edu/search.php?query=big+bang

Here is one link in particular that addresses your question, although you might not find the answer satisfactory (as it is that we really don't know).

http://curious.astro.cornell.edu/question.php?number=541

---ooOoo---

Anonymous
USA

Thank you for your message. Your question is a very important one. Admittedly, the answers that are usually provided at the level of popular science are not always precise or even right.

The evolution of the Universe at the largest scales can be described by Einstein's General Theory of Relativity. According to General Relativity, and based on what we observe today, the Universe was increasingly smaller, denser, and hotter at earlier times. Extrapolating this trend back into very early time (roughly 14 billion years ago), we reach a stage when all known laws of physics, including General Relativity, break down. This (called a "singularity" in technical terms) is what we call the Big Bang.

Therefore, the Big Bang refers to a moment that we cannot describe based on the known physics. We can understand the physics almost immediately after the Big Bang, but not at the Big Bang (i.e., the "singularity"). In order to do this, we need an extension of General Relativity, which is the focus of intensive theoretical efforts at the moment. Once we have a mathematically consistent extension of General Relativity, then we can address the question of "what happened before the Big Bang?" in a meaningful way.

As for the conservation of energy, you are right that all physical processes conserve the total energy in the Universe. However, the energy need not be all in the matter we are made of. Part (or, at some earlier time, all of it) of the energy could be in other particles that are predicted by modern physical theories. The energy can be transferred from these particles to matter and vice versa. In fact, this is done at the Large Hadron Collider (LHC) where protons are accelerated to such high energies needed to produce the Higgs Particle.

I hope you find these comments helpful.

Best Regards,

---ooOoo---

Professor Mitchell Begelman
Chair: Department of Astrophysical & Planetary Sciences
University of Colorado at Boulder
USA

Thanks for your message. This really is outside my areas of expertise, but my understanding is that this is a fundamentally unsolved problem. Guesses as to how it might be resolved suppose either that the conservation of matter and energy does not apply near the time of the Big Bang, due to physical laws which are not yet known, or that our Universe is the result of an unstable quantum fluctuation that occurred in a previously existing Universe, in which case matter and energy could have "leaked" into the Universe from a pre-existing region that is now unobservable. (This latter is the "multiverse" or Baby Universe idea.) Of course, the latter idea doesn't really solve the problem; it merely pushes it back to before the Big Bang.

You may have heard about the claimed discovery of the B-mode polarization of the Cosmic Microwave Background that was announced a few weeks ago. If confirmed, these data could help to narrow the possible explanations.

Regards,

---ooOoo---

Origins: Before the Big Bang

Professor Martin Kruczenski
Department of Physics and Astronomy
Purdue University
USA

Your question of what happened before the Big Bang is one that certainly has fascinated many physicists, but, unfortunately we don't know the answer. It is one of the big unanswered questions of modern physics. Having said that, let me just add a few things.

The most obvious possibilities are that

1. There was nothing, as you say.
2. The Universe was large, contracted and then expanded again.
3. Our Universe is an expanding bubble within another Universe.

Physics is an experimental science, so the main question is how one can find out. At the moment, there is strong experimental evidence that the Big Bang happened, namely that the Universe was, at certain time, extremely small. The more detailed knowledge comes from studying the Cosmic Microwave Background (CMB) radiation. The Universe is full of a (very weak) bath of radiation produced in the early days of the Big Bang.

Such radiation has fluctuations that reflect perturbations that existed at the very beginning of the Universe, and has confirmed the idea of the Big Bang. However, with the experiments that are done now, it seems very unlikely that the question of what happened before the Big Bang can be answered in the near future.

The other possibility is to resort to theoretical studies to see if any of those scenarios is more plausible than the others. These are the type of questions that you ask:

[Quoted from the original mail: *Given that energy and matter cannot be created, how was all the original matter and energy of the Universe created in the first place? That is, before the Big Bang?*

The reason I'm asking is that I wondered why it existed at all? And in that form? And for how long? And why did it suddenly change? (I have heard people say things like "There was nothing before the Big Bang, because time did not exist before it" or "The laws of physics were different then," or "it just is," but these sound like cop-outs to me.)]

Unfortunately, as I said, the evidence indicates that the Universe was initially very small, much smaller than anything we can see in the laboratory. When distances are so small, gravity becomes a very strong force, stronger than all the others. The problem is that there is no theoretical understanding of what happens when gravity becomes that strong. The most promising theory to answer that is called String Theory. Unfortunately, String Theory has not made enough progress to be able to even speculate on that.

The short answer, then, is that not only we don't know the answer, but that we also have very few ideas on how to answer that experimentally or theoretically. Obviously, this has happened before in physics with other problems, and, with time and hard work, people were able to figure out the answer. We'll see what happens in this case; as they say, sometimes it's about the journey...

Best,

---ooOoo---

Professor Phillip Duxbury
Chairperson: Department of Physics and Astronomy
Michigan State University
USA

There are no accepted theories for what happened before the Big Bang. Membrane theories have some speculations, but that's about it.

---ooOoo---

Professor Matt Griffin
Head: School of Physics and Astronomy
Cardiff University
UK

I'm afraid I can't provide a solution to the questions you ask. I don't know anyone who can, and I don't believe there is anyone who can.

Your questions all boil down to "Why is there something rather than nothing?" and "If the Universe emerged from nothing, why did that happen, and how can it have been nothing if it had the property that something could emerge from it?"

Our current understanding of physics breaks down in the early stages of the Big Bang, when quantum mechanics and General Relativity are not compatible. A resolution of that, and anything beyond speculation as to what happened before, will have to await new a physical theory. Even then, we may still be no closer to understanding why there is something rather than nothing.

It's interesting to speculate though. An engaging book along those lines that you might find interesting is *Why Does the World Exist?: An Existential Detective Story* by Jim Holt (Holt 2013).

Regards,

---ooOoo---

Professor Lawrence Krauss
Foundation Professor: School of Earth and Space Exploration
Arizona State University
USA

The total energy of our Universe, including gravitational energy is likely to be precisely zero. Thus there is no impediment to creating matter in the Universe from nothing. I wrote a whole book about it. *A Universe from Nothing.* (Krauss 2013)

---ooOoo---

Origins: Before the Big Bang

Professor Omar Almaini
School of Physics & Astronomy
University of Nottingham
UK

I am not an expert in Cosmology, but I will do my best to answer your questions.

In energy terms, the total energy of the Universe may actually be zero. Gravitational potential energy is negative, and in the flat Universe we appear to live in, this may precisely cancel out the positive energy of all the matter, etc. Therefore, it is possible that our Universe is just a chance fluctuation made out of nothing. Again, this is largely a speculative (though very interesting) idea, and it certainly doesn't explain where all the laws of physics came from in the first place. These ideas are explored in an accessible form here:
http://www.youtube.com/watch?v=vwzbU0bGOdc

and there is interesting further discussion here (Carroll, *A Universe from Nothing?*):
http://www.preposterousUniverse.com/blog/2012/04/28/a-Universe-from-nothing/

As for what happened "before" the Big Bang, again we do not know. It is possible that time began at this point, as you will often read in popular explanations, but we have no direct evidence of this. The evidence that the Big Bang itself occurred is overwhelmingly strong, and we can trace back the origins of the Universe to just a fraction of a second after the Universe began. Before this point, however, our laws of physics break down, so we can only speculate.

If you want to explore more of these ideas I can recommend other books/articles written by Sean Carroll. Here are some good examples:

- *Does the Universe Need God?*
  http://preposterousUniverse.com/writings/dtung/
- *How Did the Universe Start?*
  http://www.preposterousUniverse.com/blog/2007/04/27/how-did-the-Universe-start/

Best wishes,

Professor Björn Garbrecht
Theoretical Physics of the Early Universe
Physik Department T70
Technical University Munich
Germany

The explanation for the creation of matter in the Universe that most particle physicists follow nowadays, and which I also consider as very plausible, is Inflation. Field Theory and General Relativity naturally include the possibility of "vacuum energy," which has the peculiar property that it can have negative pressure. This implies that a volume filled with vacuum energy can gain energy by expanding. In contrast, the matter that we encounter in everyday life has positive pressure - meaning that, by expanding a volume filled with gas, we can gain mechanical energy by taking energy from the gas. This is, of course, what happens in combustion or steam engines. Ordinary matter, as we know it, could be created when the vacuum energy decays.

Besides creating matter, Inflation offers explanations for a few more puzzles:

- It leads to an accelerated expansion, during which the Universe has expanded by an enormous factor of at least 10^{20}. This explains why the Universe today can be so large, and how it can consist of regions that were out of causal contact after inflation.
- The expansion efficiently evades all spatial curvature, leading to the prediction that the Universe is almost spatially flat today. This is essential for the Universe to reach an age of about 14 billion years, as it is observed today. Universes that are not flat would either diverge or collapse too fast.
- Finally, while Inflation in classical General Relativity leads to a perfectly homogeneous Universe, we observe small inhomogeneities on large scales. On "small" scales, these small inhomogeneities have collapsed, due to gravity, into galaxies. These inhomogeneities emerge naturally during Inflation, when including quantum effects.

The picture described here has emerged during the past 20 years or so. The most important observational inputs come from the Cosmic Microwave Background (CMB) (WMAP and Planck experiments), and surveys of galaxy clusters that measure the distribution of matter throughout the Universe. A spectacular confirmation of inflation may be around the corner, if the B-mode polarisation in the CMB that has been observed recently by the BICEP collaboration can be confirmed to be of cosmological origin. (If not, unfortunately, the knowledge gained is limited, as this does not rule out Inflation).

As for the beginning of time, General Relativity can describe the Universe starting in a singularity (i.e. where its size is infinitely small and the energy density is infinite). However, the laws of Particle Physics break down in such a situation, such that a fair statement here is that one simply does not know what really happens close to a singularity. To avoid this embarrassment, some people suggest that the Universe undergoes an eternal cascade of phases of Inflation, during which, by the negative pressure, matter (or energy) is created from "nothing." These questions are highly interesting, but they remain highly speculative so far, in contrast to what has been learnt from the CMB and galaxy cluster surveys.

I am not really following who is currently best at conveying these exciting findings and ideas to the general public, but I hope that these remarks help you to find some excellent sources for further reading.

Best regards,

---ooOoo---

Professor Shaul Hanany
School of Physics and Astronomy
University of Minnesota
USA

Currently, physicists do not have concrete evidence about the time preceding the Big Bang. So the simple answer is: we don't know. Some physicists have theories about the time preceding the Bang, but, currently, there is no observational evidence to refute or confirm any of these theories, so they can be considered "educated speculations."

Regards,

Professor James Brooks
Chair: Department of Physics
Florida State University
USA

I think we all still have some of the same questions!

I will try to tell you briefly what we do know, and why, as we work our way back to the Big Bang, but a description of the object at the Big Bang, and what was there before, still baffles us all.

I do solid state physics, but, from my interactions with our Astro and High Energy groups, what I know is that, based on pure scientific observations, they know the Universe is expanding, and, just by playing the data in reverse, it extrapolates back in both time and dimension to the some pinpoint at the Big Bang. The Astro people can now look back very close to this event with various kinds of telescopes (especially radio telescopes), and the High Energy people can create in large accelerators (like the Large Hadron Collider in Europe that recently found evidence for the Higgs Particle), energies and temperatures that were, in effect, moments after the Big Bang (although only in tiny elementary particle interactions involving protons, etc.).

The theoretical scientists look at this data, and it all makes a very complete and consistent story going back just after the Big Bang, but what caused it is still a huge mystery, and even after the Big Bang there are still many mysteries and unsolved problems: dark matter, dark energy, the full implications, behavior and types of Higgs particles, etc.

You mention conservation of energy, and since $E=mc^2$, I think if all the energy (including dark) and matter (again, including dark) is added up in the Universe, it is conserved and accounted for now, and as far back as we can look. In fact, this controls the rate of expansion of the Universe. Likewise, we know there was a sudden change (called Inflation) that started the expansion, but, again, what triggered it is presently beyond our comprehension.

One of the interesting things is that information can only travel at the speed of light which is fixed in all situations (thanks to Einstein!), so, when we look out in deep space, we can only "see" things that occurred after the Big Bang, and that limits the distance out and the time back that we can look.

So, there could be other Big Bangs and Universes out there that we simply cannot detect. So, people start to talk about Parallel Universes, worm holes, Strings, and all sorts of exotic things. Some of this may be crazy, but we are always surprised at what is discovered, so we try to keep an open mind.

So, now I am going to pass this email on to my colleagues who really understand what is going on, and they can correct my mistakes and give you more insight.

I hope they will email you soon.

Thanks for your interest in this fundamental problem! We, in the Department, are trying to figure out the answers, and that is what makes our job fun!

Regards,

---ooOoo---

Professor Karl Ludwig
Chair: Department of Physics
Boston University
USA

I'm a solid state experimentalist, not a cosmologist. My understanding is that the questions you ask are part of ongoing research in cosmology. While there are theoretical ideas out there, none is currently regarded as an accepted answer. Moreover, some worry that there will never be a definitive answer to questions of what caused or preceded the Big Bang for lack of physical evidence.

Regarding time: As you know, time and matter are coupled by General Relativity - I think this is the origin of the idea that time itself began with the Big Bang. However, when the Universe was at the Planck length, we don't know the relevant physics, so I don't think this is a clear-cut issue. It's very difficult to think clearly about time, and our views of "cop-out" answers may simply be products of our preconceptions. Even in a traditional deistic conception of the Universe, there are usually unresolved questions about time, such as whether God would be constrained to exist in time as humans are, in which case God would be less fundamental than time itself.

Regarding mass/energy: there are interactions between space-time dynamics of the expanding Universe and mass/energy density that may, I believe, lead to changes in observed mass/energy. As you probably know, some theorists speculate that the Universe is a quantum fluctuation in which the observable mass/energy is balanced by the negative potential energy of gravity to give a zero net gy. However, currently, all of this is simply theory/speculation.

Anyway, those are fun questions, but that's about all I know.

Regards,

---ooOoo---

Prof. Dr. Uli Katz
Chair: Experimental Physics (Astroparticle Physics)
Friedrich Alexander Universität Erlangen Nürnberg
Germany

This is a very good question indeed, to which there is no confirmed, generally accepted answer (at least to my knowledge).

After collecting a little feedback from colleagues, I can confirm that my initial statement is correct: there is no generally recognised concept to explain the Big Bang as such or how it occurred, i.e. what was before the Big Bang. Also, there are no observations or experimental measurements that point to an answer to this question.

The basic problem is that the Big Bang starts in a so-called singularity, where the whole Universe is concentrated in one point. This initial state also marks the beginning of time. This description is a sound mathematical concept and a solution of Einstein's equations, but it is inadequate for describing physical reality. The main reason for this is that in the "vicinity" of the Big Bang, our two most successful theoretical concepts - General Relativity and quantum physics - are inconsistent with each other. Quantum theory does not include effects of General Relativity and General Relativity does not include quantum effects. Finding the correct theory of quantum gravity is one of the most central and, definitely, one of the most difficult questions of current research in theoretical physics.

There is a nice article entitled *Avoiding the big bang* at:
http://www.einstein-online.info/spotlights/avoiding_the_big_bang/?set_language=en
that indicates a few ideas how one could overcome this impasse.

Please note that, indeed, in cosmology, energy in its usual definition[*] is not conserved, but a function of time. A gravitational term arising from General Relativity needs to be added to "standard" energy to arrive at a new quantity that is actually conserved.

Hope this helps a little?

Best regards,

---ooOoo---

Assoc. Prof. Murray Hamilton
Physics Department
The University of Adelaide
Australia

The short answer to your question is that nobody really knows.

Our physical theories are based on observations that we make here and now, in a world that we can more or less interpret. That is to say, we can, to some degree, relate our theoretical concepts and calculations to things we experience. When we try to extrapolate our ideas to earlier and earlier epochs, we run into conundrums such as you bring up.

Nevertheless, when we do find a way to measure something that the theory tells us should be related to what happened as far back as we can go, we find that the theory is giving the correct prediction. Finding these "somethings" to measure is very challenging, and making the measurements is difficult, so progress is slow. Nevertheless a recent example of such progress was the discovery that the microwave background radiation is polarised (e.g.:
http://kipac.stanford.edu/kipac/projects/bicep2).

However, none of this makes it any easier to interpret what was happening back then, because this was a time and place (well, may-

---

[*] which you might write as $E = mc^2$ + kinetic energy + radiation energy (photons)

be!) that literally was unimaginable to us. The theory is only a crutch for our imagination, but it is the best one we have.

To get a better answer, you probably ought to direct this question to someone who specialises in cosmology. In Australia your best bets in finding one are at Math Sciences at Monash, or Physics at ANU. There is a number of astronomers at places like University of NSW or Swinburn University who might also be able to offer an opinion.

Best regards,

---ooOoo---

Alex Flournoy
Teaching Professor, Department of Physics
Colorado School of Mines
USA

In short, our understanding of the Big Bang does not imply anything "before" the initial singularity began to expand. What I mean by this is that, no physicist worth his weight would imply that we came from nothing. What the equations of General Relativity imply in a cosmological context (that is, treating the contents of the entire Universe as a fluid of radiation, matter, vacuum energy, etc.) is that its current expansion was preceded by a "beginning" of infinite density.

There are two important points to note here:

1. No one can claim that this initial singularity itself came from nothing. Perhaps it was around for eternity and simply chose the moment of the Big Bang to release itself.

2. The Theory of General Relativity, and all that it predicts, is limited by the fact that it is a classical theory. While certain theorems seem to imply that gravitational singularities are inevitable in General Relativity (e.g. black holes and the Big Bang), the theory itself is incapable of describing the physics very close to these singularities. The problem here is that, when the system becomes as small like near a singularity, its quantum mechanical behavior cannot be ignored. To fully answer the

questions of near-singular systems requires a quantum theory of gravity. We are working on that now with String Theory, loop quantum gravity and a few other candidates. But, even with all of the progress made, our understanding of these theories is far from complete. Perhaps, one day with enough development of these quantum gravity theories, we will be better equipped to answer the question of how and why a Big Bang.

---ooOoo---

Professor Peter Hoeflich
Department of Physics
Florida State University
USA

Allow me to add an addition to the response of Prof. Brooks from the view of Astrophysics and Cosmology.

Questions like yours are at the center of modern physics, and lead us to the limits of our current understanding of space and time. Because we are at the limit of our current understanding, allow me to go "step by step" in the answer of your question.

Finding the right question is intimately connected to the answer, or a better physical model beyond General Relativity (GR) and Quantum Mechanics (QM). The first theory, GR, describes the gravitation or, better, structure of the space itself on large scales, based on a space-time continuum.

GR also formulated the equivalent of mass and energy, $E=mc^2$. The second theory, QM, describes the physics of small scales and time, and is very successful in describing elementary particles and their nature.

On large scales, we see not just large distances, but also back in time (because of the limit on the speed of light), which gives us the information, and, thus, we can probe the early phases of the Universe right after the Big Bang by looking at large distances (i.e. when the light travelled some 13.7 billion light years.)

Modern astrophysics and cosmology allows us to see the imprint of the Big Bang, and the following picture is emerging by the almost isotropic micro-wave background radiation, the fluctuations and the formation of galaxy, and structure we see in the current Universe with the distribution and kind of matter/energy, the dark matter and dark energy which only have been discovered during the last 30 and 15 years respectively. (Nobel Prize in Physics, 2011).

Within GR, we understand the distribution of Galaxies, density contrasts in energy and matter of 1,000,000,000,000,000,000,000,000, 000,000,000,000,,000,000,000,000,000,000,000,000,000   by   the growth from density - fluctuations of initially 0.000001.

During the first 10^{-35} seconds (i.e. 0.00000000000000000000 00000000001 sec.), the Universe undergoes a very rapid expansion by about 300,000 light years. This, the so-called Inflation, is needed to explain the uniformity of the so-called microwave background, a background radiation discovered by George Smoot who got the Nobel Prize only some 10 years ago.

This rapid phase of expansion is needed to explain why all directions in the Universe could "communicate" with each other within the constant speed of light. Only some 3 weeks ago, BICEP, an American Telescope system at the South Pole, gave first proof of this super-rapid expansion phase (though other experiments need to confirm this discovery).

The basic theory beyond describing this early phase is General Relativity. You may ask, why can we expand within 1^{-35} seconds by 300,000 years in the first place without violation of the Einstein's Special Relativity (SR) or, better, the limited speed of light? Is GR in contradiction to SR? Does Einstein contradict himself?

The answer is "No: The constant speed of light applies to the local, non-accelerated space. During the space-time itself is rapidly accelerated and not an inertial system (frame)."

We had this problem in the past (some 30 years ago), when we saw material jets flowing from the center of galaxies with an apparent speed of light larger than c. This puzzle was solved by using the proper transformation between local system and us, the global observer (called Lorentz Transformation, a mathematical description suggested by Poincare years before Einstein gave it a meaning by the interpretation in his SR).

In the early Universe, GR and SR are right, but the Universe itself is rapidly accelerating, and we can use GR and SR to describe the Inflation. Only for an observer from the outside (us some 13.7 billion light years away in time and 50 billion light years away in space), Inflation appears as a paradox, but, still, GR allows us to describe the physics of space-time.

Your question on the conservation of energy (including mass) points towards global conservation, and GR allows a description of the Universe without violation.

Your second question on the "before" brings us to the limit of modern physics. "Before" is defined by our daily experience. Einstein, with the GR, showed that we need to consider space and time simultaneously.

Next stop: Quantum Mechanics (QM) describes small scales and the forces of electro-mechanics, weak and strong fields. GR describes the 4th force, the Gravitation.

QM shows that space and time coordinates cannot be described simultaneously with infinite accuracy. The uncertainty in time, delta t, and space/energy, delta (E) is given by the Heisenberg Uncertainty Relation which can be written as delta (T) delta (E) > h or delta (x) delta (p) > h, where h is the Planck constant which, albeit small at 1-^{27} CGS, is not zero.

As such, the meaning of time and space used in Einstein's theory breaks down at subatomic scales. As such, our description of the very early Universe requires a theory which combines QM with GR and can apply to all 4 forces of nature.

The current goal of Physics is to develop a theory to unify all forces of Nature into a combined description. Nature, during the last 10-20 years (or last 3 weeks, if you count BICEP) helps us to be guide to this goal.

In the early part of the last century, the paradox of isotropic speed of light and the nature of elementary particles and the question why stars shine lead us to a new phase of science. (Before, the world was regarded as "mechanical").

Currently, we are in a similar revolution in Physics, and we are working hard to open the new door.

These are, indeed, exciting times!

Best wishes,

---ooOoo---

Professor John Blondin
Head: College of Physics
North Carolina State University
USA

This is an excellent question, and not one that can be adequately addressed by our current understanding of the Universe.

Ultimately, our understanding breaks down at a very early time in our Universe when the density was so high that quantum effects must have dominated the gravitational space-time. This is a regime where humans do not have the ability to experiment. Without the ability to test theories, there is little chance for scientific progress.

One recent test of our understanding of the early Universe came from the reported polarization of fluctuations in the cosmic background. This effect was predicted by Inflation theory - something that I didn't expect would ever see an observational confirmation. So there is hope for progress.

You can find many theories on the web about extra dimensions, multiple Universes, etc., but all they do is push your question off to some unknown dimension.

If you want to read more about this, I recommend digging through the following blog

http://www.preposterousUniverse.com/blog/

Cheers,

---ooOoo---

Thomas Van Riet
Department of Physics and Astronomy
Catholic University of Leuven
Belgium

Your question has no definitive answer. If we would understand this, then that would be major news. At this moment in time, the physics community has no true satisfying answer. That is why we do research. All of the questions we have regarding the early Universe are difficult since we have no direct experimental guidance. Purely theoretical consistent theories are also notoriously difficult.

I think there are two options.

Either there was indeed a genuine Big Bang, in the sense that something was created out of nothing. This is indeed very puzzling. I will just give you two references to read that could help you.

The first reference is about the concept of conservation of energy, which is often not well understood. *Why and how energy is not conserved in cosmology.* http://motls.blogspot.be/2010/08/why-and-how-energy-is-not-conserved-in.html?m=1 .

Stephen Hawking has some ideas on how "natural" it could be to create something out of nothing. (Hartle & Hawking 1983; Hawking 1984)

A second option is that the Universe is eternal. It always existed. The process of Cosmological Inflation is, then, such that, to an observer, the Universe seems as to have started out of a Big Bang. But, what looks as a Big Bang was simply an exponential blow up of a piece of the Universe. Just Google Eternal Inflation, and you will find thousands of links. Then there are also ideas that suggest the Universe had a shrinking phase and then expanded (Ekpyrotic scenario), but the last option is in tension with recent observations.

In general, a Big Bang means a singularity of space-time. Probably singularities do not exist in a true quantum theory of gravity. A simple example is in String Theory: you can find a solution of a Universe that is about to undergo a Big Crunch (the time reversal to a Big Bang), but when it gets smaller than the so-called "string length" there is some new physics that effectively makes the Universe open up again. See for instance (Brandenberger & Vafa 1989).

To summarize: we have no answer to your question, just theoretical clues at the moment. BUT it could be useful to study the links I gave and further links about the concept of energy conservation in curved spaces....

Best regards

---ooOoo—

-

Prof. Dr. Thomas Posch
Institute of Astronomy
University of Vienna
Austria

Let me try to give a simple, but, hopefully, not simplistic answer.

First of all, one must ask for the perspective from which an answer is supposed to be given. Every scientific (as opposed to religious or philosophical) account of the world has its limits of validity / applicability.

Newtonian Mechanics turned out not to work fine at speeds close to the speed of light, for example. Geometrical optics will not explain the phenomena of interference of light waves, etc.

Having said this, let me turn to your questions: *"Given that energy and matter cannot be created, how was all the original matter and energy of the Universe created in the first place? That is, before the Big Bang?"*

Now, our present cosmological theory, if applied to the first fractions of seconds after the "Big Bang," is already pushed close to the limits of its applicability (i.e. applicability of Einstein's General Theory of Relativity). But if we try to go back further, beyond/before the "Big Bang," I'd say we're then clearly going beyond the applicability of this theory. And this makes not much sense for a cosmologist (note that I'm not, myself, a cosmologist, but an astronomer and philosopher).

Another question is whether the Big Bang Theory will stand the test of time on the long run.

And still another question is: Which philosophical theories might help to find answers / conceptual frameworks in the context of "Big Bang and beyond?"

This is all I can say for now.

Kind regards,

---ooOoo---

Origins: Before the Big Bang

Professor Martin Grant
Dean of Science
McGill University
Canada

I guess the first answer is that the closer we get to the Big Bang, the hotter it gets, and we certainly get to the point where gravity and quantum mechanics must be reconciled. Right now they're not.

A promising idea which might reconcile them is String Theory, which relies much more heavily on mathematics than most theories in physics. There is some very nice work on String Theory being done at McGill by, for example, Professors Keshav Dasgupta, Alex Maloney, and Johannes Walcher.

Any observational evidence for the predictions of String Theory would likely come from Astronomy (as is the case for the theory of Cosmological Inflation, which occurs at somewhat lower temperatures). That's what people mean by "the laws of physics were different then": it is conceivable that something happens to the law of conservation of mass, just as 100 years ago with the discovery of Relativity, that law had to be modified to account for $E=mc^2$. I disagree with the sentiments that there was nothing before the Big Bang, or time did not exist then, or it just is: how would people know these things?

And, in fact, I have heard a few talks on pre-Big-Bang physics. While most of them start with something like, "Let's assume..." which leaves me a little cold, there is some serious work being done on it. The phrase to Google is Trans-Planckian Physics. Some thoughtful work on potential astronomically-observable implications of trans-Planckian physics is being done by Professor Robert Brandenberger at McGill.

Cheers,

---ooOoo---

Professor Glenn Starkman
Department of Physics
Case Western Reserve University
USA

First, energy is not conserved in General Relativity. Something called "stress-energy," which is a combination of energy and pressure, is conserved, but this does allow energy to emerge "out of nothing." This doesn't explain where all the matter came from, because in the world around us, the number of "baryons" (protons + neutrons - anti-protons - anti-neutrons) IS conserved; but it turns out that that is rather of an accident and almost certainly was not the case in the early Universe.

Neither of those explains where the Universe itself came from. That is still a matter mostly of speculation, which means to say that there are theoretical ideas that emerge from particle physics and General Relativity, but that none of them are on firm enough ground that I, at least, would be willing to assert an explanation.

Cheers,

---ooOoo---

Professor Eric F. Bell
Director: Michigan Institute for Research in Astrophysics
Department of Astronomy
University of Michigan
USA

Sadly, cop-outs are the order of the day (or that's my understanding, at least). Before the Big Bang is outside of space and time, and, therefore, out of the reach of physics. There are speculations about multiverses, etc., that are well-motivated; these, at some level, are (my perspective) currently trying to make sense of something that is really beyond science.

So your questions are perfectly reasonable, but, sadly, unanswerable.

Sorry,

---ooOoo---

Origins: Before the Big Bang

Professor Ali Eskandarian
Dean & Professor of Physics College of Professional Studies &
Virginia Science and Technology Campus,
The George Washington University
USA

Our understanding of the Universe is far from comprehensive, and certainly not complete. Therefore, we tend to extrapolate based on the known laws, and our extrapolations often do not withstand the test of further experimentation. However, without posing difficult questions and pushing our current understanding, we would not be able to make progress. What makes thinking about science interesting is that we don't presume we are right. In fact, we should be prepared to accept that we are wrong most of the time, yet continue to think and pose newer and more sophisticated questions. With the advance of knowledge, some of our previously unanswerable questions find a framework to be posed and to be better-defined.

For example, before Einstein's General Theory of Relativity, elucidating the relationship between the space-time geometry and gravitational dynamics, certain questions in cosmology (and some might say even the field of cosmology itself) remained ill-defined or even ill-conceived. Since then, we have made tremendous progress, because we were able to address some of those questions within a logical and workable framework. I suspect we will modify our views about some of the questions you are posing now after we have made more progress in discovering and learning the yet-unknown laws of nature.

It is important to realize that our very concept of physical quantities such as mass, energy, time, space, etc., are based on the current state of the Universe, more or less, and our understanding of it. Some of the quantities that we take for granted now, may not have a natural definition (or as well-defined a notion) in a drastically different type of environment. Hence, you may hear statements by physicists that sound unfamiliar or contradictory to what we assume to be well-understood in our daily experience.

All of this is to say that the question you have posed as "simple," may not be as simple as you think it is. Indeed, deeper questions would require deeper understanding, which we certainly don't possess when it comes to addressing questions regarding the origins of our Universe. The short answer to your question is that I don't know enough to be able to give you a satisfactory explanation without having to resort to theoretical constructs that would only complicate the matter without adding to the understanding.

I am, however, taking the liberty of copying Professor Parke, who has made substantial contributions to our understanding of the laws of nature (including my own way of thinking). If anyone, he would be able to shed more light on the questions you have posed. I do hope you will continue exploring and thinking about deeper questions.

With best wishes,

---ooOoo---

Professor Bill Parke
Professor of Physics and Chairman of the Department of Physics
Columbian School of Arts & Sciences
The George Washington University
USA

By selecting Prof. Ali Eskandarian, you could not have picked a better person to ask your questions. He is one of the most profound thinkers about fundamental questions, with minimal encumbrance of prejudicial thoughts or personal aggrandizement of any I know about.

His explanation of our current difficulty in answering the questions you pose is spot on.

As you, I also do not find it useful to dismiss our lack of understanding as a reflection of impossibility or a predisposition that we cannot explain. Rather, past experience has shown that careful thought deepens our models of how things work. Moreover, if we do not search both new ideas and the frontiers of observation, we are almost certainly not going to progress.

I can say a little about your question on mass and energy. In our current understanding, these quantities cannot be properly measured until you can "stand away from them." The best way (in our current theories) is to measure their gravitational pull. Now, if you have to be some distance from a localized region of matter and energy to measure them, then they cannot be defined for the whole of the Universe, since the Universe is everything which can be measured, and therefore there is no outside to stand and observe.

Now, apart from measurement, some of our current theories about the Universe start with our Universe confined to a space smaller than the nucleus of an atom, with no matter, but just strong gravity and high pressure to expand. Material was created in the condensation during expansion. Such an Inflation Theory, consistent with Einstein's General Relativity Theory, predicts a relic set of gravitational waves, whose effects may have recently been observed.

Cheers,

---ooOoo---

Professor Steven Furlanetto
Professor of Physics & Astronomy
University of California at Los Angeles
USA

You ask a very difficult question! Unfortunately, the answers that you feel are "cop-outs" are probably the most accurate…there is simply nothing that science can say for sure on the topic of the pre-Big Bang Universe.

That said, there is one theory that is quite attractive and gives an intuitive picture of how things might unfold. This is called chaotic or eternal inflation, and you can read a good summary at the Wikipedia article *Eternal Inflation* ((Wikipedia 2014f).

Another alternative is to dispense with the Big Bang entirely – see the at the Wikipedia article *Cyclic Model* (Wikipedia 2014e).

Basically, though, this changes the question from "what happened before the Big Bang?" to "what does it mean for the Universe to last forever?" At this point in our scientific understanding, there are simply questions that aren't answerable!

Best,

---ooOoo---

Origins: Before the Big Bang

Dr. Brian Espey
Physics Department
Trinity College
Ireland

Your question is beyond a scientific answer currently, I'm afraid, as our physics knowledge breaks down at the high energies and small size of the early Big Bang, as we really need a fully unified "Theory of Everything" in order to be able to address it. What we can do is use physics to extrapolate very close to the start, and make some informed guesses beyond that.

With regards to time, the Newtonian concept of absolute time is not believed any more, and the idea of "before" and "time" is a moot point when there is no Universe (sorry!). As to where things sprang from, recent work shows that the structures of the current Universe (galaxies and clusters etc.) sprang from quantum fluctuations (generation of particles and anti-particles that always goes on at the small scale) that then got stretched out into larger units... but then we don't understand why exactly there was a slight imbalance between particles and anti-particles (to the tune of one in 100,000,000 or so, based on the ratio of photons in the Universe to the number of particles), other than to say that the symmetry was not perfect... which goes to show that ideas of what is "perfect" and "beautiful" need to give way to the facts of reality.

I hope that this answers your questions to at least some degree. You can be sure that the things that you are thinking about also exercise some of the greatest minds including Roger Penrose and Stephen Hawking!

Best wishes,

---ooOoo---

Anonymous
USA

The origins of things are the most difficult questions, and, in my view, some are unknowable. The answers you list are speculation. But the Inflationary Model predicts creation of Universe out of "vacuum," i.e. no ordinary matter and energy, but some other field, and seems to work. But what was before then, etc., we do not have a solid model for.

---ooOoo---

Professor Douglas Hamilton
Department of Astronomy
University of Maryland
USA

I'm afraid that I do not have easy answers for you - you are asking about questions at the very edge of what science can accomplish. We have absolutely no information (or way of obtaining information) from before the Big Bang.

Nevertheless, scientists like to try to understand anyway! Here is an excellent article by Steve Nadis (Nadis 2013) that addresses your question directly - although it probably raises more questions in the process! Happy Reading!

Cheers,

---ooOoo---

Professor Joseph Dwyer
Head: Department of Physics and Space Sciences
Florida Institute of Technology
USA

No one knows the answer to your question.

Physics can tell us what happened back to a very small fraction of a second after the Big Bang, but, so far, we don't know what, if anything, happened before. Some theories envision our Universe as kind of a bubble in a sea of many bubbles, with bubbles giving birth to more and so on. This doesn't entirely answer the question, since one might then ask where did all that come from? Since science can only deal with things we can observe and measure, these questions may need to be answered by philosophers and not physicists.

Hope this helps.

Best Regards,

---ooOoo---

Professor Daniel Kay
Professor of Physics
Department of Physics
University of British Columbia
Canada

You wrote ".... I wondered why it existed at all." That's the deepest mystery, the mystery of existence. Why is there something rather than nothing? The short answer is that nobody knows. I don't know if the question is even answerable.

You asked about the origin of matter and energy. One idea is that the total energy of the Universe is exactly zero. That is possible because gravitational (potential) energy is negative. (That is related to gravity being always attractive.) It's a neat idea, the creation of the Universe from nothing. However, it is only speculation, and no one knows how it would work. (Could we discover laws which determine how Universes are created?)

You mention the statement "There was nothing before the Big Bang, because time did not exist before it." That kind of statement has never made any sense to me either. The most widely held view today, among cosmologists and particle physicists, is that our observable Universe is part of an infinite "multiverse." Within this multi-multiverse, "Big Bangs" are happening all the time, creating other "Universes". If String Theory is correct (and it also is just speculation), then the laws of physics can be different in different regions. This is appealing because we wouldn't have to explain "why these laws?". Whatever laws operate in our part of the multiverse would have to be such as to allow for our existence.

Even though so much is speculation, most of it is nevertheless based on our current understanding of things. It is noteworthy that we are now in a position to attempt to try to explain the origin of the observable Universe as a physical event. (Of course, it still wouldn't answer the ultimate question.) Whenever you hear or read about any scientific ideas, try to find out the observational or experimental evidence. Then you can make a judgement about how speculative it is.

I hope that has been helpful.

Sincerely,

---ooOoo---

Prof. Dr. Luciano Rezzolla
Institute for Theoretical Physics
Albert Einstein Institute, Am Muehlenberg
Germany

Our best gravitational theory (General Relativity) breaks down when trying to explain what happens at the Big Bang. It predicts that the Universe was contained in a singularity, i.e. a single point with infinite density, and that expanded after that. This is mathematically possible, but physically unrealistic. A theory of quantum gravity that describes this process has been sought for the last 50 years without success so far.

Note, however, that the fact that the whole Universe was concentrated in the singularity does not require any "creation." Think of a box (the singularity) which contained matter at very large density that "opened," and that released all the matter that was before concentrated. Time starts when the singularity expands (the box opens). There is no time in a single point, and, hence, the time can be thought to have started when the box opened.

Best wishes,

---ooOoo---

Professor Sun Kwok
Chair Professor of Physics and Dean of Science
Department of Physics
University of Hong Kong
China

Science is limited to what can be observed, and any theory has to be able to be tested. What was there before the Big Bang is beyond what science can do and cannot be answered by the scientific method.

---ooOoo---

Professor Sara Ellison
Physics and Astronomy
University of Victoria
Canada

The short answer is that we don't know! Although we have theories in which we are fairly confident down to fractions of a second after the Big Bang, what happened before, or why it happened remains a mystery. It is a regime about which we can ponder (and certainly there are theories concerning this epoch), but robust scientific theories require empirical data to support them. And we have no observations that probe "before the Big Bang," and no real prospects for doing so. For now, it remains one of the Universe's many mysteries!

Regards,

---ooOoo---

Origins: Before the Big Bang

Professor John Learned
Department of Physics and Astronomy
University of Hawaii at Manoa
USA

Well, this is, of course, a deep question. In the current cosmology, the standard story is that it is all something from nothing. Time did not exist before the Big Bang. The whole Universe is a huge fluctuation.

Indeed, there have been models which involve cyclical Universes, though all information would be wiped out between cycles. The perhaps depressing standard model take now is that we face a long and cold expansion to isolation and cold death many billions of years in the future.

In the standard view, the question of before t = 0 has no meaning, as unintuitive as that may be. Nobody much likes this, but other models of an infinite and renewing Universe simply do not work.

It is a strange place in which we live...

(In any event you and I will be long gone! For now, enjoy what we have and leave as many good things as you can.)

---ooOoo---

Professor Pao-Ti Chang
Professor of Physics
Department of Physics
National Taiwan University
Taiwan

You are actually asking a difficult question.

Based on the current understanding of the Universe and physics laws, the matter and energy already existed at the beginning of time. Although there are physicists who are considering the physics before the Big Bang, most scientists do not discuss it. The reason is simple. We don't have information and the laws of physics should be understood based on the facts or information.

We know that time is absolute zero, and space is tiny in the very beginning according to Einstein's General Relativity, and many experimental facts, such as cosmic micro wave background.

There is no time before the Big Bang or creation; therefore, "how long before creation" may be meaningless for physicists. Scientists have their limitations. I know that I have not provided you with a satisfactory answer, but this is what I understand.

---ooOoo---

Anonymous
UK

Thanks for the email. The answer is that it can and probably did come from nothing.

Have a look at this article by Steve Nadis (Nadis 2013), and, in particular, the section entitled "Something from Nothing."

You may also find this article about Stephen Hawking of interest (Barrick 2011). The idea of something from nothing doesn't make sense, but only on a large-scale - what we term the "Newtonian" world - i.e. the world covered by Newton's Laws.

At very small scales, quantum mechanics takes over. Similarly at high speeds, Relativity takes over. Hard to see why time would appear to slow down for objects moving close to the speed of light - but it does, and has been proved experimentally.

I would add that, in terms of the details of all this, you'd need to ask someone else - I am not a particle physicist or cosmologist. My suggestion, if you wanted to follow it up, would be to look at the Cambridge Applied Mathematics and Theoretical Physics webpage, look at staff/grad student profiles and perhaps contact one of them. (Not Prof Hawking).

Final point - and this is not a cop-out, honestly! - the theory behind all of this requires very, very complex mathematics (well beyond me) and is hence very hard to explain.

All the best

---ooOoo---

Professor Avishai Dekel
Professor: Department of Physics
Hebrew University of Jerusalem
Israel

The current situation is that we don't have answers to these questions within the realm of physics.

The reason is that we are still missing a physical understanding of what happens when both strong gravity and quantum effects are important, as is valid close to the time of the "Big Bang."

This is the missing theory of quantum gravity, the search for which started with Einstein and has not been completed yet. Without it, we do not really understand the Big Bang, and cannot address it in a scientific way. It is a "singularity" point in space time. We don't know what happened before it, and whether time had any meaning then. We don't even know what questions to ask. The only thing we know is that all observations are consistent with a Universe that is very, very dense everywhere and expanding very rapidly at a time that is a small fraction of a second after the time we call "the Big Bang."

Regards,

---ooOoo---

Anonymous
Norway

The answer is simple:
we do not know.

Different scientists might have their own pet theory for that, but there is no consensus and no definite answer. This is one of the big open questions of modern cosmology (and physics). It is also a very difficult one to address, for the simple fact that we do not have the tools to address it.

First, physics, as we know it, ceases to work at the Planck scale which is after the Big Bang. Second, we can't easily make observations. So, you can imagine how difficult it may be to make progress.

In other words, cosmology and physics and astrophysics have made huge progress in the past 10-20 years, but many deep questions remain unanswered, and one of them is "what powered the Big Bang? What was the physics under those extreme conditions?" Which is deeply related to the issue of how we can make gravity and quantum mechanics to agree (as they are, they don't).

In summary, you are not the only one being puzzled, and believe, me you are in good company.

Best,

---ooOoo---

Professor Mark Birkinshaw
Professor of Cosmology and Astrophysics
School of Physics
University of Bristol
UK

The general idea is that the total energy content of the Universe is close to zero, so no energy or matter were necessarily "created" in the Big Bang. Instead, there was a conversion from an undifferentiated state (some sort of quantum field) into a differentiated state after the field decayed into particles (positive mass-energy) with relative gravitational binding (negative potential energy) and relativistic motions (positive kinetic energy). The sum of these energies then equals the energy of the field which decayed.

Now, the value of that original field energy density is not known. It could be precisely zero (i.e., the Universe is a sort of ultimate free lunch), or could be a local excess of energy in a larger Universe surrounding our visible Universe ... that is, on a large enough scale, the Universe could have structure which is not visible within the small region (only about 15 billion light-years across) that we can study. In this picture, our visible Universe can be thought of as "detaching" from the larger entity. This is the general idea of "Chaotic Inflation."

The moment, around 15 billion years ago, that we identify as the Big Bang would then be the moment that the quantum field within the larger Universe randomly, and locally, decayed into the state which could differentiate into particles and their motions. Within that larger Universe, there would then be regions that did something different - and at many different times. So, our visible Universe would be one chunk of a larger entity which happens to be hospitable to the type of intelligence that can ask questions about what caused the Universe.

You might find one of the (many) books talking about "Inflationary Cosmology" to be helpful if you'd like to go into this in more detail - there are a lot of ideas in the area, and a few books which try to explain it. Guth came up with part of the idea, and wrote a book "The Inflationary Universe: the quest for a new theory of cosmic origins," (Guth 1998) which might be a decent start, but the picture we tend to use more, these days, is that of Andrei Linde.

I hope this helps.

---ooOoo---

Professor Graham Woan
Professor of Astrophysics
School of Physics and Astronomy
University of Glasgow
UK

Thanks for your query. It's not my specific field, but I think the honest bottom-line answer is (as Patrick Moore would say) "we simply don't know." In fact, we don't know whether the question you are asking is even a sensible question, though it sounds reasonable on the face of it. To start with, it's not a given that mass-energy cannot be created. In fact, one of the most popular theories ("Inflation") that describes much of what we actually see in the early Universe includes the generation of mass as a consequence of the generation of spacetime.

My own guess is that a deeper understanding, based on observation, will probably reveal that the question you are asking is just not appropriate, and is based on a rather local perspective on the Universe. As an analogy, someone who is learning about latitude and longitude on the Earth, but has only local knowledge of the ideas of north/south/east/west might be bothered about what happens at the north pole. They might say it's a cop-out when they are told, "There's nothing there to impede your movement, but you can't go north from there, just south." For them, this would conflict with their long-held understanding of what moving n/s/e/w means, but for us, it would just indicate that their understanding of coordinates was skewed by local experience.

Despite all appearances, substantial astronomy (in the form of astrophysics) is only about 60 years old and younger than, say, quantum mechanics. We are learning fast, but there's still a long way to go I'm afraid! Sorry if that's not a very satisfactory answer, but I (like many astronomers) take the views of theoretical cosmologist with a big grain of salt, and concentrate on gathering observational evidence for what's going on.

Best wishes,

---ooOoo---

Professor Tom Marsh
Professor of Astronomy
Department of Physics
University of Warwick
UK

I am afraid I will have to go for another answer you may consider to be a cop-out, which is, I don't personally think anyone can answer this at the moment.

As I understand it, in the current theory, "before the Big Bang" is not a legitimate question (there are an infinite number of such questions possible in science: e.g. "what atom has -1 protons in its nucleus?"), not that people do not sometimes consider it (there are theories in which the Big Bang is one of many).

In the case of the Big Bang however, it may well only be not legitimate because of the inadequacy of current theory. It is well known that General Relativity, the theory that describes the expansion of the Universe, is incompatible with quantum mechanics that controls microscopic physics. This normally does not matter, but certainly does early in the Universe when quantum effects on a Universal scale matter.

As one gets very close to the beginning, it is thus we are sure that our current theory, as standardly applied, fails. (It is also the case that the conditions very early on are not directly testable - we can't make experiments at high enough energies - so it is a tough area to progress.)

By the way, you may have heard of "dark energy" - there is good evidence for this, which means that the energy content of the Universe has continued to grow ever since the Bing Bang, and is still growing today - the dark energy is already dominant and will dominate ever more. Personally, I don't think we have much of an idea about this either. It's good to have things we don't understand.

That's my pennyworth, but I am not an expert, and will hand you on to [a colleague] who teaches cosmology here and certainly knows more than I do, and can probably shoot me down in flames,

Regards,

---ooOoo---

Anonymous
USA

These questions are really great questions.

There is no way to answers all of them in a short e-mail.

Very briefly: We do *not* know if the laws of physics are the same when approaching the beginning of the Universe. But all current observations or experiments confirm that conservation of energy (in its general relativistic formulation) is preserved. We have almost no understanding about what was before Big Bang.

I believe you will find the book *The First Three Minutes* by Weinberg (Weinberg 1993) useful.

Best,

---ooOoo---

Professor Heidi Jo Newberg
Professor of Physics, Applied Physics and Astronomy
Rensselaer Polytechnic Institute
USA

Nobody knows what happened before the Big Bang, or at the instant of the Big Bang. Our known equations of physics are known to not apply at or before the beginning of the Universe. That is because General Relativity and quantum mechanics are not consistent with each other at very small volumes or very high densities. There must *be* laws of physics that govern that situation, but we do not know what they are.

Some people think that Universes can grow out of other Universes - suddenly popping into existence like a new bud on a flowering bush. Others think Universes could be created spontaneously out of nothing. Others defer to God. So, the answers you have gotten before roughly cover our knowledge of the answer to that question.

I hope this helps.
Best Wishes,

---ooOoo---

Professor Jean-Pierre Caillault
Professor of Astronomy
University of Georgia
USA

I don't want to pretend I'm an expert on this topic, so, instead, let me refer you to an article that quotes someone (Sean Carroll) who is:
http://www.Universetoday.com/2008/06/13/thinking-about-time-before-the-big-bang/
Good luck!

---ooOoo---

Origins: Before the Big Bang

Origins: Before the Big Bang

Professor Vincent Rodgers
Professor, Nuclear and Particle Theory Group
Department of Physics and Astronomy
The University of Iowa
USA

The short answer to your question is that it is unknown what happens, but there is no reason to believe that it is impossible to understand, at least in the future. Theoretical physics is a research area just as medicinal chemistry, and we need input in order to advance the field. That input can be from experiments such as Hubble's Galaxy Recessions or more fundamental problems at the mathematical level that String Theory addresses. I've attached three papers by Fr. Lemaitre (Lemaître 1931b; Lemaître 1931a; Lemaitre 1934), one of the important founders of modern day cosmology, in order to strip the magic and mystery from this conversation that has crept into colloquial conversations regarding cosmology.

The Big-Bang (or, as Lemaitre called it, the Primordial Atom) is a solution to Einstein's field equations for a gravitational ric. There are 10 non-linear partial differential equations for the gravitational field, and at least four more for the matter content in the Universe (or wherever you want to apply these equations). There is no known "method" for solving non-linear partial differential equations. However, if one makes the assumptions that the Universe is rather democratic with its distribution of matter (homogeneous), then the field equations become quite tractable and reduce to a family of solvable ordinary differential equations.

This is what Friedmann showed in 1922, but died not long ter. Fr. Lemaitre redid these calculations and applied the idea to Hubble's data. There are strong consequences that *can* be measured if these assumptions about the boundary (or initial conditions) are met.

Recent research with BICEP II, COBE, WMAP, and the ESA Planck missions, show remarkable agreement with these assumptions, *and* they also show at what scale these assumptions breakdown.

With that said, Lemaitre's calculation, and their present day embellishments, seem to be beautifully correct, but do not probe the region of space-time that you are discussing. There, many issues arise, and theoretical physics has only recently been able to even write down a mathematical theory for gravitation in such an arena (String Theory). It is still very much in infancy in theoretical physics, so no discussion is reporting any acceptable scientific understanding of that region.

Other issues that arise are that Einstein's equations are differential equations, which means they require calculus in three spatial dimensions and one time dimension, i.e. four dimensions. However, it is now known that four dimensional manifolds can support not one distinct way of doing calculus, but an unaccountably infinite number of ways of calculus (see Donaldson's invariants). However, quantum field theory, which will be at least the minimal framework for this region, does not need calculus, so it will include all such "differential structures."

But what this quantum theory is, and how we write a mathematical law that is consistent with other known principles, is under investigation. All the laws of physics you mention, related to mass and energy, are precise mathematical expression in Einstein's theory and are not assumed. They are consequences of strict mathematical consistencies.

With that said, one can ask a different question, "What is an electron made of?", and also stir up lively conversations about its constituents and have debates just like the initial Big Bang. But, the fact of the matter is, we do not know, and, although it might be a fundamental particle by today's standards, it might not be by tomorrow's standards. Remarkably, Quantum Field Theory is mathematically consistent, and makes incredibly accurate predictions about elementary matter assuming that the electron is fundamental. This 11 decimal place accuracy with today's experiments is sure to give way to better and better experiments as we carefully continue to investigate our Universe.

Best,

---ooOoo---

Professor Richard Creswick
Department of Physics and Astronomy
University of South Carolina
USA

While I am not a cosmologist, I think it is safe to say that you have hit on several really difficult questions that, as far as I know, have no satisfactory explanation yet.

You are correct that matter/energy cannot be created or destroyed, and one of the really important questions in cosmology is "Why is there more matter than antimatter in the Universe?" All the processes we know transform energy into equal amounts of matter and antimatter.

We are pretty confident we can follow the evolution of the Universe from a few seconds after the Big Bang to the present. The problem with going back further is that gravity begins to play a really dominant role, and, so far, no one has come up with a good theory of Quantum Gravity.

Hope this helps,
Yours,

---ooOoo---

Anonymous
Taiwan

The question you asked is a deep question. Nobody in the world knows the answer, but looking for the answer is what physicists are doing. I am afraid I cannot give you a better answer.

Regards

---ooOoo---

Anonymous
Canada

I think that you would be best off by tackling one of the many popular books about the Big Bang (e.g. *Before the Big Bang* by Brian Clegg (Clegg 2011)).

The answers to your questions, as far as they are known, are bound up in Quantum Gravity, String Theory and other ideas that may be relevant before the "Planck Time," and will be a little difficult to capture in email - and far less satisfying.

Best wishes,

---ooOoo---

Professor Jochum van der Bij
Physikalisches Institut
Albert-Ludwigs-Universität Freiburg
Germany

I will try to answer your questions in a, hopefully, understandable way.

The answers will not be fully correct; for that you would have to study a lot of physics.

Your letter actually contains two logically separate questions. One is where the matter comes from, and the other is what happened before the Big Bang (i.e. before time.)

The first question is, at least qualitatively, and also largely quantitatively, understood. When we go back in time far enough, there is not nothing, but still strong spacetime curvature plus a so-called inflation-field, which may be an effective description of Quantum Gravitational effects. The details of this are the subject of research, and one expects the Planck measurements to give more information.

The presence of the inflation field leads to a very fast expansion of the Universe. This is because the field contains "energy," whereby the meaning of energy in this context is somewhat subtle. Due to the dynamic interactions between the curvature and the inflation field, the inflation field will start to oscillate after a certain time. These oscillations create the ordinary matter, including the dark matter, which is considered ordinary in this context.

After this, the inflation stops, and the expansion of the Universe is slowed down. All normal processes, like the formation of galaxies, the cooling of the background radiation, etc. then follow known laws of physics. This kind of physics is principally well understood, whereby the main questions are, what particles are precisely there and what is the precise form of the inflation, etc.

The other question is what happened "before" all this. Of course this is an old question, already considered by Augustinus, which was put in the form: "Was the Universe created in time or was time created with the Universe?"

All indications from modern science are that time and the Universe were created together. Within Relativity, time and space exist only in the overall form of curved geometry, and there is no such thing as an absolute time. When one tries to apply Quantum Mechanics to General Relativity, the notion of geometry disappears; in this sense, time, and also space, is only a coarse-grained description on large scales.

Presumably, the dynamics takes a discrete character, where one can only correlate physical quantities among each other, like the value of some field and the curvature of spacetime. This, presumably, in a probabilistic way.

Time, in this sense, is only a relational time. It is not a fundamental object of the theory. It is not known what precise mathematical structures play a role here. The subject is Quantum Geometry. This is a mathematically very complicated subject. Unfortunately, experimental, and even cosmological information, appears to be absent. However, there are a number of approaches, that give sensible results, but none of them are definite at the moment, and one cannot decide among them.

This is not really my field, but I contributed a little. You may find my podcast at the Freiburg colloquium of interest. ("Did God have a choice when he created the world?" (Einstein). (http://podcasts.uni-freiburg.de/mathematik-physik-biologie/physik/2011/71619737)

I hope this is of some help to you.

Best wishes,

---ooOoo---

Prof. Dr. Matthias Bartelmann
Zentrum für Astronomie
Universität Heidelberg
Germany

You raise a very interesting and intriguing question here. The honest answer is that we don't know.

It is well possible that the Big Bang was an event in the evolution of the Universe that was preceded by some other physical state, but we have currently no means to test ideas and speculations about the possible evolution of the Universe prior to the Big Bang. The main reason is that we don't know how to combine the Theory of Gravity with the Unified Theory of everything else in physics, but we would need a combined theory to develop physically well-founded concepts about epochs possibly preceding the Big Bang.

It is, however, well possible to create matter and energy. This was demonstrated by Fred Hoyle in the 1950s, and also Albert Einstein seems to have had similar thoughts.

I'm not sure whether you find my answer satisfactory, but I believe that at the moment we can only admit to our ignorance.

All the best,

---ooOoo---

Professor Michael M De Robertis
Department of Physics & Astronomy
York University
Canada

Thanks for the questions.

It's important to realize that science works on the basis of models; i.e., formulating an idea (normally in mathematical language), testing the idea, and then modifying it if the results of the test aren't entirely consistent with the original model. Scientists also like to keep asking "how" and "why" questions; they get annoyed if they reach a "brick wall."

The Big Bang appears to be a phenomenally successful model for the Universe. It has its limitations; it is completely unable to describe our Universe when it was younger than about 10^{-43} seconds. For this, we will require a model/theory that employs Quantum Gravity, something that isn't yet on the horizon in physics. Yet, even if a theory of Quantum Gravity is formulated, it will still only be able to describe the Universe as a model.

You are right in my view, however, to wonder where all the mass-energy came from. Some cosmologists now speak about the "multiverse," an infinite ensemble of Universes in which only a small number of Universes have parameters consistent with intelligent life. Even if there is such a thing as the "multiverse" from which our Universe emerged about 13.8 billion years ago, the multiverse still would owe its existence to an "earlier" configuration of mass-energy (e.g., a quantum field of pure energy). And so forth.

It seems to me that there are only three possibilities.

- First, one is led to the (classical) infinite regress, where our Universe depends on an earlier Universe which depends on an even earlier Universe, etc. This is no explanation at all.
- Second, one is led to postulate the creation of the first Universe (or master Universe) from which every other Universe emerged through time.
- Third, and this is becoming even more popular, one could simple state that the Universe has always been and does not require a creator or first cause. While one may ask about cause and effect for things *in* the Universe, one is not entitled to ask about an origin or cause *of* the Universe itself. In other words, the third possibility actually assigns to the Universe a property that theists assign to God; "necessary being."

I find only the second possibility plausible, myself, though there are an amazing number of details to work out. This is good news for scientists, of course.

Anyway, not sure whether I've been of any help here.
Regards,

---ooOoo---

Professor Geraint Lewis
School of Physics
University of Sydney
Australia

Firstly, conservation of energy does not always hold. Einstein's General Relativity famously does not conserve energy when space-time expands – we see this in the Universe as photons lose energy as the Universe expands. So, you can extract or lose energy in the Universe.

But, secondly, we simply don't know the process that brought the Universe into existence (i.e. we don't have the maths that allow Gravitation and Quantum Physics to work together), so we don't know what came before. So, the energy that exists in the Big Bang could have existed before the birth of the Universe, or could have been produced when a previous Universe collapsed. The phrases you mentioned are not really cop outs - they are basically statements that we don't know.

Cheers

---ooOoo---

Professor Amy Barger
Department of Astronomy
University of Wisconsin-Madison
USA

I just heard Prof. Alex Filippenko from UC Berkeley give a great talk on this very topic at the University of Hawaii's Kennedy Theatre. (His lecture should soon appear on the Institute for Astronomy, University of Hawaii web site.) In the meantime, you can read this short excerpt from his book (Filippenko & Pasachoff 2001) at: http://www.astrosociety.org/publications/a-Universe-from-nothing/

or ask him directly at AskAlex.org (a new website he announced at his talk that currently looks very basic!)

Best wishes,

---ooOoo---

Origins: Before the Big Bang

Professor Bradley Peterson
Chair: Department of Astronomy
Ohio State University
USA

I'm sorry for the long delay in responding, but it's been the end of the academic year with lots to do and also I'm traveling. In fact, I'm writing to you from Bern, Switzerland, just a few steps away from where Albert Einstein developed his special theory of relativity in 1905.

The simple, honest answer is that no one can really answer any of your questions with authority. Dark Matter is the easiest, as the evidence for it is overwhelming: it does have gravitational effects, but otherwise hardly interacts with normal matter. Dark Energy is hard to understand: essentially it means that the energy density of the Universe remains constant: as the Universe expands, new energy somehow shows up. Again, that's what best (i.e., most consistently) explains how we observe the Universe to be expanding. We are now in a state where the Universe is beginning to inflate more rapidly than it should in the absence of Dark Energy: this happened once before, when the Universe was very young, but then suddenly stopped. Will this rapid expansion phase stop too? I don't know.

As for what might have existed before the Big Bang, there is one theory that predicts that the signature of gravitational waves could survive the Big Bang (i.e., we're not completely out of touch with whatever happened before the Big Bang). This is kind of what the controversial BICEP2 results are about (in the news lately, but whether the observations are secure is being debated).

Anyhow, we're not there yet: we still have fundamental things we don't understand - we used to call them "metaphysical" questions, as they are beyond the reach of normal physics. But the truth is, we've made remarkable progress in understanding the Universe over the last 100 years, and it shouldn't surprise or upset us that we don't understand it all yet. Undoubtedly, even some of the things we think we understand will turn out to be incorrect. Science is really hard...

Hope this is helpful if not completely satisfying...

Regards,

---ooOoo---

Origins: Before the Big Bang

Anonymous
UK

This is a difficult question to answer: essentially "what happened before the Big Bang?" The answer is: nobody knows. We don't know the physics that preceded the Big Bang; no observations that we can currently make can probe to this time. So, anything is really just guess work. Latest observational results an certainly probe very early times in the Universe, but nothing from "before" the Big Bang.

Of course, there are theories, but at the present time that's all they are: ideas. There are some reasonable articles online that discuss some of these (eg, http://science.howstuffworks.com/dictionary/astronomy-terms/before-big-bang.htm or http://www.hawking.org.uk/the-beginning-of-time.html and similar pages).

So, even though it may seem like a cop-out, I'm afraid that that is all the answer that I have, even if it may appear unsatisfactory!

Best wishes,

---ooOoo---

Professor Joseph Moody
Department of Physics and Astronomy
Brigham Young University
USA

Every theory has parts that are assumed with no deducted understanding of exactly why they should be that way. For the Big Bang Theory, it is that all matter emerged as energy at a specific time in the past.

We claim no understanding as to why that should have been the case. All the explanations you have heard *are* cop-outs. We only assert that it happened and ascribe to no theory as to what the universal time and space were like before that event. It may hurt the pride of scientists to admit it, but speculations on what came before are only speculations and not really science.

Regards,

---ooOoo---

Professor Georges Meylan
Directeur: Laboratoire d'astrophysique
Ecole Polytechnique Fédérale de Lausanne (EPFL)
Switzerland

We know, with some confidence, what happened a few minutes after the Big Bang, through the results of the transmutation of light elements, such as hydrogen and helium. As we go back closer to the Big Bang, our models are more and more uncertain, and then fail completely because of the lack of a theory of gravitation in the context of quantum physics.

Before the Big Bang, we do not know anything, and we do not even know if the question is meaningful, since the Big Bang marks the beginning of time and space. Our present understand of these remote periods of time are unsatisfactory, and shall, hopefully, be the subject of significant progress in the coming decades.

Best regards,

---ooOoo---

Professor David Hobill
Department of Physics and Astronomy
University of Calgary
Canada

The issue of what happened "before" the Big Bang, and even "at" the Big Bang, is impossible to answer, since the only theory that consistently accounts for the dynamics of our Universe is General Relativity, and that theory breaks down at the moment when the Big Bang occurred. It leads to a singular solution, where the structure of spacetime, energy, density, the fields in the spacetime - everything has an infinite value. This means that energy conservation - conservation laws of any form - break down. Mathematically, infinity is an undefined quantity. Physically, an infinite quantity cannot be used to make any definite predictions.

Currently, there are no theories that provide alternative explanations. General Relativity is a classical theory. (Newtonian cosmology also predicts an initial infinite Big Bang.) What is needed is a

Quantum Theory of Gravity which should be able to describe what happened at and, perhaps, before the Big Bang. This is much like what happened when classical electromagnetic theory has no way of explaining the structure of the electron (and it still cannot do so). The electron, before quantum theory arrived, was described by a singular point (with an infinite value for the electric field) in space and time.

Many people asked the question: "What is the electron made of?" But, since the mathematical quantities (electric field, charge density, electric potential, etc.) were all infinite, there was no way to explain what an electron consisted of until quantum field theory was able to describe an extended object made of quantum electric fields that did not involve infinite quantities.

Thus, we are in the same situation with the Big Bang. A theory that allows us to sort of rid Classical Gravity Theory of its singular values will likely provide the answers, and that is one of the reasons for the ongoing research in developing a theory of Quantum Gravity.

While there have been many descriptions of what happened at the Big Bang and before, none are based upon theory. They are at best philosophy, perhaps poetry, but not physics, since physics requires a theory rooted in a mathematical description - that is to say, simple logic. Mathematics is developed simply to encode logical statements at a higher level than going back to simple truth tables - just as computer programming languages encode the binary logic of off-on switches that is the basis of digital computation, communication, etc.

Therefore, no one can say what happened before or at the Big Bang – the mathematics breaks down and, therefore, so do any logical inferences.

Hope this helps explain why there is no description of what happened or what the conditions were before or at the Big Bang that allowed our Universe to form.

If you still have any questions about this let me know.

All the best,

---ooOoo---

Origins: Before the Big Bang

Professor J Richard Gott
Department of Astrophysical Sciences
Princeton University
USA

I've discussed these things at length in my book *Time Travel in Einstein's Universe* (Gott 2002). I would recommend that. It's on Amazon.

In General Relativity, there is local conservation of mass-energy - that is, the energy and mass in this room can only go up if something or someone comes in the door. But there is not a global conservation of mass energy in cosmology where there is no place at infinity that is flat to take an energy standard.

All the photons in the Universe lose energy as they redshift as the Universe expands, for example. And in Inflation, the negative pressure in the vacuum state means that the energy in the vacuum stays constant as the volume of the Universe increases.

General Relativity has passed many experimental tests. So has Inflation.

Anyway, I had a model with Li-Xin Li, where the Universe was its own mother. Inflationary Universes can give rise to other Inflationary Universes, like branches growing from a tree, Linde showed. We said, "What if one of the branches curls back around and grows up to be the trunk?" Then the beginning of the spacetime looks like the numeral 6 with a small time loop at the bottom. In this model, every event would have events that preceded it, but the Universe would have a finite beginning. The earth is finite to the east, but there is no easternmost point. In our Universe, the Universe was finite to the past but had no earliest event.

Others propose the Universe quantum tunnels from a state with no geometry.

Anyway, I have explained all these ideas in *Time Travel in Einstein's Universe*.

---ooOoo---

Professor Peng Oh
Department of Physics
University of California Santa Barbara
USA

Unfortunately, I'm going to reply with the same cop out - we just don't know. No one has managed to come up with an answer that is testable, and our known laws of physics (like General Relativity, which is used to describe spacetime) break down then. It is a question for future generations...

Good luck!

---ooOoo---

Professor Ming Chung Chu
Department of Physics
Chinese University of Hong Kong
China

I don't think there's an answer to your question from science yet. The standard scenario is the following:

1. Before Big Bang (or Hot Big Bang), there was an Inflationary phase, during which the Universe expanded by a huge factor, in an exponential growth manner. This Inflationary growth was driven by a huge potential energy which was somehow there.

2. This Inflationary phase existed for very short moment, about 10^{-32}s. When it ended, the standard Hot Big Bang began, which was a much slower expansion of the Universe. At the end of the Inflationary phase, the potential energy (that drives the Inflation in the first place) got converted into kinetic and thermal energy. Some of these become matter (by the famous $E=mc^2$). So, most of the energy and matter we see today came from the huge potential energy before the Big Bang.

3. There is some observational evidence of this Inflation Theory, especially from the recent BICEP2 results of the B-mode polarization in the cosmic microwave background. However, its far from being established as the "truth."

4. Even if Inflation is true, it still would not answer your question. You would probably press on and ask: "Where did the potential energy come from?" At some point, science meets its limitation. In fact, what you hear people say about known physics laws break down, etc., honestly reflects our ignorance about the beginning of the Universe.

Hope this helps a bit, even though it's not answering your question.

Regards,

---ooOoo---

Professor Alan Calder
Department of Physics and Astronomy
Stony Brook University
USA

I agree that some of the things that are said about it sound like cop-outs, but, unfortunately, sometimes that is about all that can be said. We don't completely understand the very early Universe, and the things that we can say reflect that.

My expertise is not in cosmology or the physics of the early Universe. But I do know astronomy, and I can tell you that observations unquestionably show that the idea of the Big Bang and the expanding Universe are valid. No one has seriously questioned that in years.

There are some surprises, though. In 1998, observations of distant stellar explosions (type Ia supernovae) indicated that the rate of expansion of the Universe is increasing. This suggests a "dark energy" driving that expansion, but no one understands what that is.

As for the early Universe, think of it this way - we know that the Universe is expanding, and, as it does, it cools. So, if we imagine turning the clock backwards, we see that the early Universe had to have been very hot and dense. Observation of the cosmic microwave background, light emitted early on that we can now observe, shows this paradigm is valid. But, beyond that, we don't completely understand the physics, so it is difficult to describe in detail what is going on.

The point of experiments like those conducted at the Large Hadron Collider is to better understand high energy environments in which the normal laws of physics we experience don't apply. The people who work in this area try to build conceptual models of how the physics works, and from that, ideas of how the Universe formed follow. Experiments and tests of the physics in these regimes are difficult, so progress is slow, and, hence, the less than strong statements (if not cop-outs) about the formation of the Universe. One thing to note is that "Whys" of all this are certainly very hard to answer.

Hope this helps.

This Wikipedia article on the early Universe seems reasonable: http://en.wikipedia.org/wiki/Chronology_of_the_Universe

---ooOoo---

Professor Stacy McGaugh
Department of Astronomy
Case Western Reserve University
USA

I don't think there are satisfactory answers to these questions.

Our present knowledge takes us back to a tiny fraction of a second after the start of the expansion of the Universe known as the Big Bang. We do not have the capacity to say with certainty what happened all the way to t = 0, let alone t < 0.

While it does seem like a cop-out, it is exactly correct to state that what happened "before" time began is, by definition, not defined. The problem is conceptual. How do we extrapolate our linear perception of time of a tiny slice of the history of the Universe to back before Time itself began? Is time really linear, or is that just our limited perspective?

I have heard some argue that the period of Inflation near the beginning of time solves this problem, since the expansion of the

Universe goes then as R ~ exp(t) so doesn't actually get to R -> 0 until t -> -infinity. I have not heard this repeated for a long time, so perhaps its proponents feel that is also a cop-out. It does answer the "be-"before zero" question, but it begs the question of what makes t=0 special? Why did the Universe transition from its previous state to its current one?

Personally, I like the idea of an oscillating Phoenix Universe; one which expands, contracts, then is reborn to expand again. But just because I find that appealing doesn't mean it has anything to do with reality.

---ooOoo---

Postscript

So, after all that, where am I?

I think I am a little closer to understanding the complexities and the uncertainties.

It is possible that something spontaneously came from nothing, but that is still too abstract for my brain to comprehend.

The oscillating universe has a nice symmetry to it – like one that contracts through a giant black hole to the singularity, and then re-emerges on the other side. Of course, that still does not explain where it all came from or why it happened.

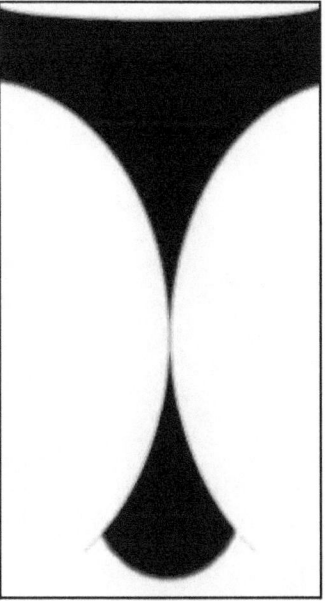

Perhaps all space, matter, energy, light and time are nothing substantial, but are merely dimensions, or properties (with many more not yet perceived) of an eternal "Now," that existed as single point, void of form, and the Big Bang was not so much of a "Bang" as a Division and Unravelling of these dimensions from the single point.

Why did they divide and unravel? Perhaps it was just time to do so.

References

Aguirre, A. (2013) How did our universe come to be? *Sky & Telescope; 60 Greatest Mysteries*, 79-82.

Barrick, A. (2011) Stephen Hawking Explains Creation, Big Bang Sans God, *The Christian Post*, <http://www.christianpost.com/news/stephen-hawking-something-out-of-nothing-is-possible-53589/> (Accessed 04/08/2014).

Barrow, J. D. (1997) *The Origin of the Universe: Science Masters Series* (Basic Books).

Bojowald, M. (2010) *Once Before Time: A Whole Story of the Universe* (Knopf).

Brandenberger, R. & Vafa, C. (1989) Superstrings in the early universe, *Nucl. Phys. B*, 316(2), 391-410; Abstract at: http://www.sciencedirect.com/science/article/pii/0550321389900370.

Clegg, B. (2011) *Before the Big Bang* (Saint Martin's Griffin).

Davies, P. (2008) *The Goldilocks Enigma* (Mariner Books).

Duncan, T. (2008) *Your Cosmic Context: An Introduction to Modern Cosmology* (Addison-Wesley).

Filippenko, A. V. & Pasachoff, J. M. (2001) *The Cosmos: Astronomy in the New Millennium* (Brooks/Cole).

Gott, J. R. (2002) *Time Travel in Einstein's Universe* (Mariner Books).

Guth (1998) *The Inflationary Universe* (Basic Books).

Hartle, J. & Hawking, S. (1983) Wave function of the Universe, *Phys. Rev. D*, 28(2960), See Abstract at: http://journals.aps.org/prd/abstract/10.1103/PhysRevD.28.2960.

Hawking, S. & Mlodinow, L. (2008) *A Briefer History of Time* (Bantam).

Hawking, S. (1984) The quantum state of the universe, *Nucl. Phys. B*, 239(1), 257-576. Abstract at: http://www.sciencedirect.com/science/article/pii/0550321384900932.

Holt, J. (2013) *Why Does the World Exist?: An Existential Detective Story* (Liveright).

Krauss, L. (2013) *A Universe from Nothing: Why There Is Something Rather Than Nothing* (Atria Books).

Lederman, L. M. & Hill, C. T. (2007) *Symmetry and the Beautiful Universe* (Prometheus Books).

Lemaitre, G. (1934) Evolution of the Expanding Universe, *Proceeding of the NAS*, 20, 12-17.

Lemaître, A. (1931a) The Expanding Universe, *Mon. Not. R. Astron. Soc.*, 91, 490-501.

Lemaître, A. (1931b) A Homogenous Universe of Constant Mass and Increasing Radius accounting for the Radial Velocity of Extra-galactic Nebulae, *Mon. Not. R. Astron. Soc.*, 91, 483-490.

Musser, G. (2014) Gravitational Waves Reveal the Universe before the Big Bang: An Interview with Physicist Gabriele Veneziano, *Sci. Am.*, Blog: <http://blogs.scientificamerican.com/critical-opalescence/2014/04/03/gravitational-waves-reveal-the-universe-before-the-big-bang-an-interview-with-physicist-gabriele-veneziano/>.

Nadis, S. (2013) What came before the Big Bang? *Discover*, <http://discovermagazine.com/2013/september/13-starting-point> (Accessed 04/08/2014).

Penrose, R. (2012) *Cycles of Time: An Extraordinary New View of the Universe* (Vintage).

Steinhardt, P. J. (2011) The Inflation Debate: Is the theory at the heart of modern cosmology deeply flawed? *Sci. Am.*(April), 36-43 http://www.physics.princeton.edu/~steinh/0411036.pdf.

Stenger, V. J. (1990) The Universe: the ultimate free lunch, *European Journal of Physics*, 11, 236-243.

Turing, A.M. (1959) Computing machinery and intelligence, *Mind*, 59, 433-460.

University of St Andrews (2009) Nobel Lecture, New Teaching Space Opening, Building Named, *University of St Andrews News Item*, https://www.st-andrews.ac.uk/physics/news/Panda_news/bds_nobelopening_30_11_09.htm.

Veneziano, G. (2004) The Myth of the Beginning of Time: Two Views of the Beginning, *Sci. Am.*(May), http://www.scientificamerican.com/article/the-myth-of-the-beginning-of-time-two-views-of-the-beginning-2004-04-26/.

Vilenkin. Alexander (2013) What Came Before the Big Bang? pp. http://discovermagazine.com/2013/september/13-starting-point.

Weinberg, S. (1993) *The First Three Minutes: A Modern View of the Origin of the Universe* (Basic Books).

Wikipedia (2014a) Absolute Zero, p. <http://en.wikipedia.org/wiki/Absolute_zero> (Accessed 31/07/2014).

Wikipedia (2014b) Big Bang, p. https://en.wikipedia.org/wiki/Big_Bang (Accessed 31/07/2014).

Wikipedia (2014c) Conformal cyclic cosmology, p. https://en.wikipedia.org/wiki/Conformal_Cyclic_Cosmology (Accessed 31/07/2014).

Wikipedia (2014d) Cycles of Time, p. https://en.wikipedia.org/wiki/Cycles_of_Time (Accessed 31/07/2014).

Wikipedia (2014e) Cyclic model, p. <http://en.wikipedia.org/wiki/Cyclic_model> (Accessed 31/07/2014).

Wikipedia (2014f) Eternal inflation, p. <http://en.wikipedia.org/wiki/Eternal_inflation> (Accessed 31/07/2014).

These pages are available for readers' notes

www.ingramcontent.com/pod-product-compliance
Lightning Source LLC
Chambersburg PA
CBHW022101170526
45157CB00004B/1431